Cambridge Lower Secondary

# Science

## STAGE 8: STUDENT'S BOOK

Mark Levesley, Gemma Young, Aidan Gill,
Beverly Rickwood, Stuart Lloyd, Sheila Tarpey,
Nigel Saunders

# Collins

William Collins' dream of knowledge for all began with the publication of his first book in 1819. A self-educated mill worker, he not only enriched millions of lives, but also founded a flourishing publishing house. Today, staying true to this spirit, Collins books are packed with inspiration, innovation and practical expertise. They place you at the centre of a world of possibility and give you exactly what you need to explore it.

Collins. Freedom to teach.

Published by Collins
An imprint of HarperCollins*Publishers*
The News Building
1 London Bridge Street
London
SE1 9GF

HarperCollins*Publishers*
1st Floor Watermarque Building
Ringsend Road
Dublin 4
Ireland

Browse the complete Collins catalogue at
**www.collins.co.uk**



10 9 8 7 6 5 4

ISBN 978-0-00-836426-7

MIX
Paper from responsible sources
FSC C007454

This book is produced from independently certified FSC™ paper to ensure responsible forest management.

For more information visit:
**www.harpercollins.co.uk/green**



British Library Cataloguing-in-Publication Data
A catalogue record for this publication is available from the British Library.

Updating and contributing authors: Mark Levesley, Gemma Young, Aidan Gill, Beverly Rickwood, Stuart Lloyd, Sheila Tarpey, Nigel Saunders
Contributing authors: Fran Eardley
Development editors: Anna Clark, Julie Thornton, Rose Parkin, Gillian Lindsey, Tony Wayte, Fiona McDonald, Sarah Binns
Product manager: Joanna Ramsay
Project manager: Amanda Harman
Content editor: Tina Pietron
Copyeditors: Debbie Oliver, Naomi MacKay
Proofreader: Elizabeth Barker
Safety checker: Joe Jefferies
Indexer: Jackie Butterley
Illustrator: Jouve India Private Limited
Cover designer: Gordon MacGilp
Cover artwork: Maria Herbert-Liew
Internal designer: Jouve India Private Limited
Typesetter: Jouve India Private Limited
Production controller: Lyndsey Rogers
Printed and bound by: Grafica Veneta in Italy

The publishers gratefully acknowledge the permission granted to reproduce the copyright material in this book. Every effort has been made to trace copyright holders and to obtain their permission for the use of copyright material. The publishers will gladly receive any information enabling them to rectify any error or omission at the first opportunity.

Test-style questions and sample answers have been written by the authors. These may not fully reflect the approach of Cambridge Assessment International Education.

# Contents


How to use this book v

## 1 Respiration and moving

1.1 Blood 3
1.2 Human respiratory system 8
1.3 Skeleton, joints and muscles 17
End of chapter review 23

## 2 Nutrition

2.1 A balanced diet 28
2.2 The effects of lifestyle on health 37
End of chapter review 45

## 3 Ecosystems

3.1 Habitats and ecosystems 49
3.2 Bioaccumulation in food chains 58
3.3 Invasive species 64
End of chapter review 69
End of stage review 73

## 4 Structure and properties of materials

4.1 Structure of an atom 79
4.2 Paper chromatography 83
End of chapter review 87

## 5 Solutions and solubility

5.1 Concentration of solutions 90
5.2 Solubility 94
End of chapter review 97

## 6 Chemical changes

6.1 Using word equations 100
6.2 Pure substances and mixtures 105
6.3 Measuring temperature changes 109
6.4 Exothermic and endothermic processes 113
6.5 The reactivity series 119
End of chapter review 126
End of stage review 129

## 7 Measuring motion

**7.1** Measuring distance and time 133
**7.2** Speed and average speed 136
**7.3** Distance/time graphs 141
End of chapter review 145

## 8 Forces

**8.1** Balanced and unbalanced forces 149
**8.2** Turning effect of a force 155
**8.3** Pressure on an area 161
**8.4** Pressure and diffusion in gases and liquids 166
End of chapter review 175

## 9 Light

**9.1** Reflection 179
**9.2** Refraction 184
**9.3** Coloured light 189
End of chapter review 196

## 10 Magnets

**10.1** Magnets and magnetic materials 200
**10.2** Electromagnets 208
End of chapter review 215
End of stage review 218

## 11 The Earth and its resources

**11.1** Climate and weather 223
**11.2** Climate change 230
**11.3** Renewable and non-renewable resources 235
End of chapter review 244

## 12 Earth in space

**12.1** Asteroids 248
**12.2** Stars and galaxies 253
End of chapter review 261
End of stage review 264

Periodic Table 266
Glossary 267
Index 275
Acknowledgements 279

# How to use this book

This book is designed to support you with the content you need to learn on your course and to provide opportunities for challenging yourself. Have a go at the more challenging questions in dark green, blue, orange or purple, and read more about the scientific world in the discovering sections.

Chapter 2 . Topic 1

## A balanced diet

You will learn:

- To identify what makes a balanced diet for humans
- To describe the functions of the nutrients needed in a balanced diet
- That carbohydrates and fats can be used as an energy store in animals
- To identify and control risks in practical work
- To choose experimental equipment and use it correctly
- To take accurate measurements and explain why this matters
- To present and interpret scientific enquiries correctly
- To describe the application of science in society, industry and research

These lists show what you will learn

### Starting point

| You should know that... | You should be able to... |
|---|---|
| Animals and humans need nutrition | Recognise some dangers when doing experiments |
| Humans eat many different plants and animals | |

This table helps remind you of what you know, and the scientific skills that you have. You will build on these as you study this topic

### Diets

Your **diet** is what you eat and drink. You need food for:

- energy
- growth and repair
- health.

A **nutrient** is any substance needed for energy or used as a raw material to make other substances. Your diet should contain the following nutrients:

- **proteins** (for growth and repair)
- **carbohydrates** (for energy)
- **fats** (also called **lipids**, and used to store energy)
- **vitamins** and **minerals** (for health, growth and repair)
- **water** (for dissolving and carrying substances around the body and for temperature control through sweating).

**Fibre** is not classed as a nutrient because it is not digested or absorbed by the body. You do need fibre in your diet

Key terms

**carbohydrate**: nutrient needed for energy. Examples include starch and sugars (such as glucose).

**diet**: what you normally eat and drink.

**fats**: nutrients needed by your body to store energy.

**lipids**: another word for fats.

**minerals**: nutrients that living organisms need in small amounts for health, growth and repair. Also called mineral salts.

28 Nutrition

You should learn the meanings of the key scientific terms in bold. You can find their meanings in the margin and in the glossary (near the end of the book)

Try the questions to check your understanding

2.1

though to keep your intestines (gut) healthy. Fibre forms most of your solid waste or **faeces**. A lack of fibre may cause **constipation**, when your intestines become blocked. Good sources of fibre include wholemeal bread, brown or wholegrain rice, cereals, lentils, nuts and fruits.

1. List *three* nutrients that your body needs.
2. a) Why is fibre not a nutrient?
   b) Why does your body need fibre?
   c) List *two* good sources of fibre in your diet.
3. Give an example of a carbohydrate.

### Key terms

**nutrient**: a substance that an organism needs to stay healthy and survive.

**proteins**: nutrients you need for growth and repair.

**vitamins**: nutrients you need in small amounts for health, growth and repair.

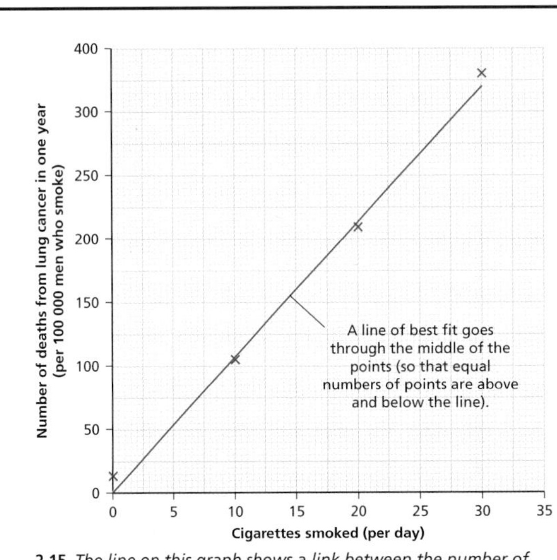


**2.15** *The line on this graph shows a link between the number of cigarettes smoked and the number of deaths from lung cancer.*

### Science in context: Smoking tobacco

A scientific study found that between 1990 and 2015, smoking caused one in every 10 deaths in the world. It also found that the number of smokers in the world is increasing, because the world population is increasing.

The percentage of people who smoke in some countries has fallen (such as Brazil, Australia and Nigeria). In others, there was no change (such as Bangladesh, Indonesia and the Philippines). The percentage of smokers in some groups has increased (such as women in Russia).

Smoking creates jobs, such as tobacco farming. Governments also get money from taxes on tobacco. However, smoking causes diseases and many countries try to reduce the number of smokers. They put warnings on cigarettes and raise the taxes on them.

Discover more about where scientific ideas have come from and how they are used around the world now

1. Explain why scientists are doing more studies on how the amount of protein in the diet affects growth.
2. Explain how a lack of calcium in the diet affects growth and development.
3. Suggest why people who do not eat enough protein are less strong than they could be.

Challenge yourself with these questions in dark green, blue, orange or purple

**2.11** *Obesity is caused by overeating and lack of exercise.*

## Obesity

As well as a lack of nutrients, an unhealthy diet can also have too much of a particular nutrient. An example is **obesity**.

If your body does not use all the carbohydrates you eat, it turns them into fats and stores them. Overeating carbohydrates and fats makes people larger. A lack of exercise makes the problem worse. If a person gains a lot of weight, they may become obese, which means that their health is in danger because of their weight.

Health problems caused by obesity include **type 2 diabetes** and **high blood pressure**.

In type 2 diabetes, too much glucose is dissolved in your blood plasma. This can damage organs, such as the heart and eyes.

### Key terms

**obesity**: being so overweight that your health is in danger.

**type 2 diabetes**: disease in which there is too much glucose in your blood, which can damage organs.

### Activity 2.2: Investigating food recommendations

What are the food recommendations in different countries?

**A1** Use different books and/or the internet to discover how much of each nutrient your country recommends. Also discover the values from a country far from yours.

**A2** Present your information as a table, to compare the values.

Try out the science for yourself with step-by-step activities

### Key facts:

- ✔ Your diet needs to contain nutrients (carbohydrates, fats, proteins, minerals and vitamins) for energy, growth and repair, and health.
- ✔ A balanced diet contains the right amount of the different nutrients from a wide variety of different foods.
- ✔ You also need water and fibre in your diet.
- ✔ Carbohydrates and fats store energy, which is measured in joules (J) or kilojoules (kJ).

### Check your skills progress:

- I can identify independent, dependent and control variables.
- I can write a risk assessment.

### Making links

In Stage 7 you may have learned about energy transfers. Describe the main energy transfers in your body.

Try the link questions to strengthen your understanding across topics

Check your mastery of key ideas and skills with this list

This section helps you check that you understand the scientific ideas and can apply them to new situations

## End of chapter review

### Quick questions

1. In your body, carbohydrates that you do not need for energy are converted into:

   **a** fats **b** proteins

   **c** sucrose **d** glucose

2. An important mineral for bones is:

   **a** sodium **b** iron

   **c** calcium **d** vitamin D

3. Another name for the trachea is the:

   **a** bronchus **b** windpipe

   **c** oesophagus **d** air sac

4. A poor diet can lead to obesity.

   **(a)** What is obesity?

   **(b)** Describe the type of diet that causes obesity.

5. **(a)** Give *one* reason why we need to eat each of these nutrients:

   - carbohydrates
   - fats
   - proteins.

   **(b)** Name *one* other nutrient that we need.

   **(c)** Describe what may happen if we do not eat enough of that nutrient.

6. Name the addictive drug in tobacco smoke.

End of chapter review **45**

## End of stage review

1. The tables show nutritional information from two different foods.

| Sliced bread | Amount per serving (1 slice) |
|---|---|
| Energy | 340 kJ |
| Fats | 4 g |
| Carbohydrates | 12 g |
| Protein | 2 g |

| Cookies | Amount per serving (1 cookie) |
|---|---|
| Energy | 330 kJ |
| Fats | 5.7 g |
| Carbohydrates | 7.2 g |
| Protein | 1 g |

(a) Cedric says that bread contains more energy than cookies. Explain why he cannot make this conclusion.

(b) What information would you need to be able to make a fair comparison between the nutrients in the foods?

(c) State what your body uses carbohydrates for.

2. The diagram shows a knee joint.

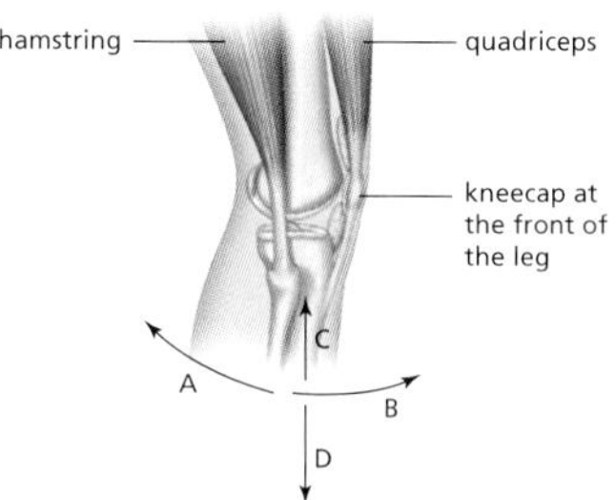


(a) When the quadriceps contracts, in which direction does it move the lower leg bones? Choose from A, B, C or D.

(b) The hamstring moves the lower leg bones in the opposite direction to the quadriceps. What is a pair of muscles like this called?

(c) Why do bones need to be moved by pairs of muscles, rather than single muscles?

End of stage review 73

At the end of the stage, try the end of stage review! This contains questions on all the chapters.

## Glossary

**absorption**: the way in which an object takes in the energy reaching its surface.

**accurate**: an accurate result is one that is close to the real answer.

**addictive**: substance that makes people feel that they must have it.

**aerobic respiration**: respiration that requires oxygen to release energy from glucose.

**alkaline**: having the properties of an alkali.

**alloy**: a mixture of metals with other elements.

**alpha particle**: particle with a positive charge given out by some radioactive elements – it is smaller than an atom.

**alveolus** (plural alveoli): tiny, pocket-shaped structure in lungs where gas exchange happens.

**analogy**: a type of model that compares something unfamiliar to something more familiar.

**angle of incidence**: this is the angle between the incident ray and the normal.

**angle of reflection**: this is the angle between the reflected ray and the normal.

**angle of refraction**: this is the angle between the refracted ray and the normal.

**anomalous results**: results which don't fit the pattern of the other results obtained.

**antagonistic pair**: two muscles that pull a bone in opposite directions.

**apparent depth**: how deep something appears to be.

**asteroid**: object composed of rock that is too small to be a planet and orbits the Sun.

**asteroid belt**: area of space between the orbits of Mars and Jupiter that contains most of the asteroids in the Solar System.

**astronomical unit**: a unit of distance equivalent to the distance between Earth and the Sun, just under 150 million km.

**attract**: pull closer together.

**average**: the mean average of a set of numbers is found by: total of all the numbers added together/how many numbers there are

**balanced diet**: eating many different foods to get the correct amounts of nutrients.

**balanced forces**: when the resultant force is zero.

**ball and socket joint**: joint where a ball-shaped piece of bone fits into a socket made by other bones.

**bioaccumulation**: build-up of a substance in an organism because the substance cannot be broken down and is not excreted.

**biofuels**: fuels made from renewable sources such as crops.

**bioplastic**: a material that can be used in the same way as plastics produced from oil, but which is produced using renewable sources.

**black hole**: a collapsed star with a very high mass, found at the centre of most galaxies.

**blood**: liquid tissue that carries substances around the body.

**blood vessels**: tube-shaped organs that carry blood around the body.

**breathing**: movements of muscles in your respiratory system that cause air to move in and out of your lungs.

**breathing rate**: the number of times you inhale and exhale in one minute.

**bronchioles**: small tubes leading from the bronchus in a lung.

**bronchus** (plural bronchi): large tube leading from the trachea into a lung.

**cancer**: when cells in a tissue start to make many copies of themselves very quickly.

**capillary**: tiny blood vessel found in all the tissues of your body.

Glossary 267

You can look up definitions for key terms in the glossary

# Biology


*Chapter 1: Respiration and moving*

1.1: Blood 3

1.2: Human respiratory system 8

1.3: Skeleton, joints and muscles 17

End of chapter review 23

*Chapter 2: Nutrition*

2.1: A balanced diet 28

2.2: The effects of lifestyle on health 37

End of chapter review 45

*Chapter 3: Ecosystems*

3.1: Habitats and ecosystems 49

3.2: Bioaccumulation in food chains 58

3.3: Invasive species 64

End of chapter review 69

End of stage review 73

1

## Chapter 1

Respiration and moving

### *What's it all about?*

During an eye test, an eye doctor looks at tiny blood vessels inside your eye. The optician may take a photo, like this one. All cells, including cells in your eyes, need a constant supply of oxygen and food from blood. Your heart pumps blood around your body by putting pressure on it. However, too much pressure can damage the delicate blood vessels in the eye. This can cause eye problems.

You will learn about:

- The parts and function of blood
- Aerobic respiration
- The parts and function of the human respiratory system
- How gases are exchanged between your blood and the air
- How muscles operate joints in the skeletal system

You will build your skills in:

- Planning investigations to test hypotheses
- Making risk assessments to identify and control risks
- Collecting and recording observations and measurements
- Evaluating experiments and investigations and suggesting improvements

Chapter 1 . Topic 1

# Blood

You will learn:

- To describe and understand the role of the elements that make up blood
- To plan investigations, including fair tests, while considering variables
- To choose experimental equipment and use it correctly
- To make predictions based on scientific knowledge and understanding
- To use results to describe the accuracy of predictions
- To carry out practical work safely
- To take accurate measurements and explain why this matters
- To describe results in terms of any trends and patterns and identify any abnormal results
- To present and interpret scientific enquiries correctly
- To describe the application of science in society, industry or research

## Starting point

| You should know that... | You should be able to... |
|---|---|
| Cells are the building blocks of organisms and can be seen using a microscope | Plan an investigation to test a hypothesis |
| Cells form tissues, which form organs and these work together in organ systems | Use a hypothesis to make predictions |
| Cells are usually specialised by having features that allow them to carry out their functions | |

## Circulation

Your heart is a pump that pushes **blood** around your body through tubes called **blood vessels**. This movement of blood is your **circulation**. Your heart, blood and blood vessels form an organ system called the **circulatory system**. It supplies all your cells with the substances they need (including oxygen and digested food substances).

## Blood vessels

Your circulatory system has two circuits. One takes blood from the heart to the lungs and back to the heart. The other takes blood from the heart to the rest of the body and back again.

The smallest blood vessels are **capillaries**. They have walls that are only one cell thick, which means that substances can pass through the walls quite easily. There are also gaps between

### Key terms

**blood**: liquid tissue that carries substances around the body.

**blood vessels**: tube-shaped organs that carry blood around the body.

**circulation**: movement of blood around the body.

**circulatory system**: group of organs that moves blood around the body.

the cells, which allow plasma to leak out of the capillaries and carry dissolved substances to all the cells in a tissue.

Altogether, your heart pumps about 5 litres of blood every minute, through about 100 000 km of blood vessels!

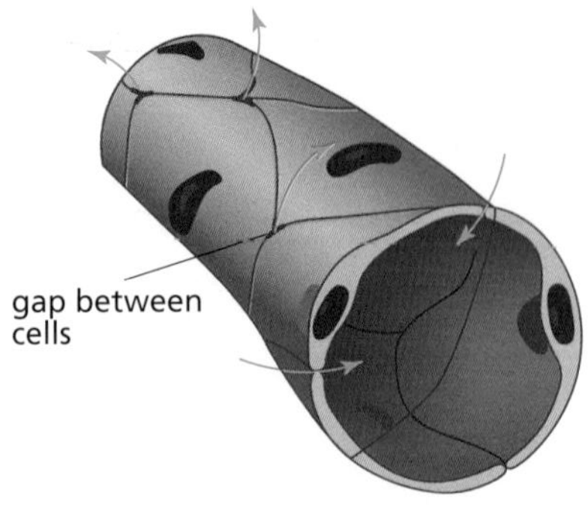


**1.1** *Capillary walls allow substances to enter and leave easily.*

## Blood

About half your blood is a pale-yellow liquid called **plasma**. It transports various different kinds of blood cells. It also contains many dissolved substances, including:

- digested food substances or **nutrients** (absorbed from the small intestine)
- carbon dioxide (a waste product of **respiration** in your cells which your lungs excrete).

**1** Describe *three* functions of plasma.

**2** From where to where does plasma carry the following substances?

**a)** carbon dioxide

**b)** nutrients

**3** **a)** Sketch a diagram to model the circulatory system. Use boxes to represent the organs and capillary networks.

**b)** Add labels to explain what your model shows.

**c)** Add labelled arrows to show *one* route taken by carbon dioxide.

### Key terms

**capillary**: tiny blood vessel found in all the tissues of your body.

**nutrient**: a substance that an organism needs to stay healthy and survive.

**plasma**: liquid part of the blood.

**respiration**: chemical process that happens inside cells to release energy.

**Red blood cells** contain **haemoglobin**. This substance collects oxygen in your lungs. As it does this, it becomes a bright red colour. The haemoglobin releases the oxygen to the cells in your tissues. Haemoglobin that is carrying less oxygen is a dull red colour.

Red blood cells do not contain nuclei or mitochondria, and so there is more space for haemoglobin. They have a dimpled shape, which gives them a large **surface area** for their size. The larger the surface area, the faster substances can enter and leave a cell.

**White blood cells** protect your body from infections caused by **pathogens** (microorganisms that cause disease). They are able to change shape easily, which lets them squeeze through the gaps in capillary walls and get into tissues. Some white blood cells wrap themselves around microorganisms and destroy them.

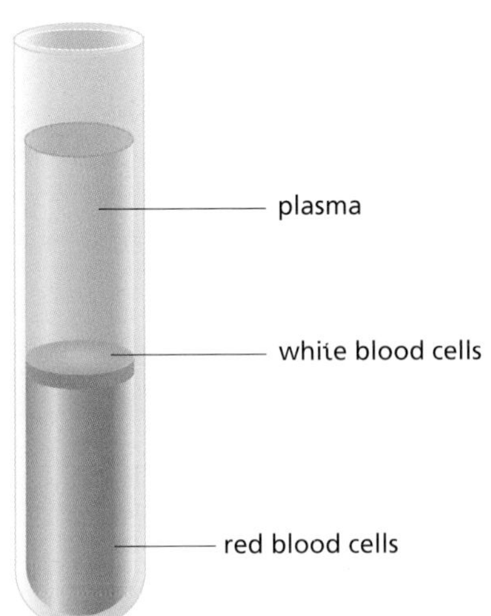


**1.2** *Blood contains many different parts.*

## Science in context: Blood transfusions

Your heart pumps blood around your body under pressure. If somebody suffers a really bad cut, this pressure causes blood to pour out of their body. With less blood in the body, the pressure is not great enough to get blood to all the important organs (such as the brain). This is why people who are seriously injured can die from loss of blood.

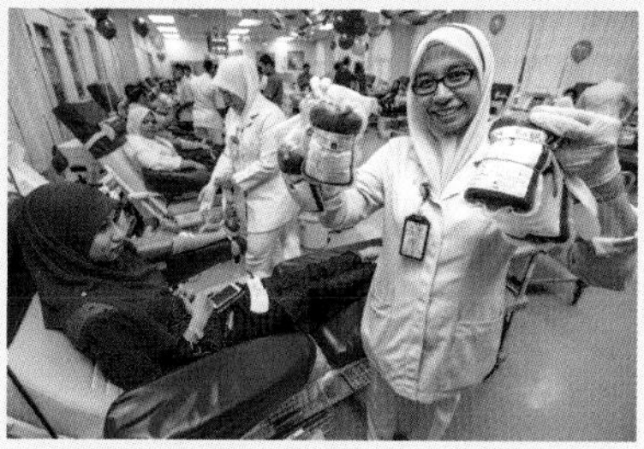

**1.4** *All donated blood needs to be clearly marked with its blood group – A, B, AB or O.*

Doctors have tried various ways to stop blood pressure falling. In the 18th century, they tried adding liquids to an injured person's circulatory system, including water, milk and other people's blood. These methods were usually not successful.

The reason why other people's blood did not always work was discovered in 1900; people have blood of different types or 'groups'. Giving the wrong blood group to someone causes serious problems. Today, if someone needs extra blood (a blood transfusion), they are given blood that matches their own blood group.

People need to donate their blood to help others. In many areas there is a shortage of donated blood or it is difficult to store. So, scientists are trying to develop artificial blood.

*Cell:* red blood cell
*Function:* carries oxygen
*Adaptations:*
- no nucleus or mitochondria so that there is more room for haemoglobin
- its indented shape increases its surface area, so that it can absorb oxygen quickly

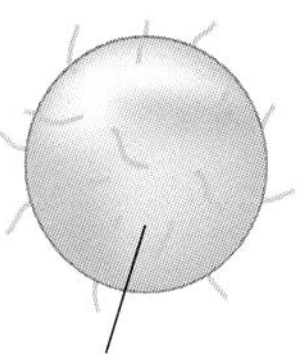

*Cell:* white blood cell
*Function:* to destroy microorganisms
*Adaptation:* flexible shape allows it to squeeze into all different parts of the body.

**1.3** *Blood cells are adapted for their functions.*

4. Which part gives blood its colour?
5. Draw a table to show what the different parts of the blood do.
6. Describe, fully, how you could use a microscope to observe some blood cells.
7. **a)** How are red blood cells adapted to contain as much haemoglobin as possible?
   **b)** Why is this useful for red blood cells?

### Key terms

**haemoglobin**: substance that traps oxygen.

**pathogen**: a microorganism that causes a disease.

**red blood cell**: cell that contains haemoglobin so it can carry oxygen.

**surface area**: the area of a surface, measured in squared units such as square centimetres ($cm^2$).

**white blood cell**: cell that helps destroy microorganisms.

**8** Compared to the number of red blood cells, how many white blood cells are in blood? Explain how you reached your answer.

**9** Although red blood cells carry oxygen, they do not need a supply of oxygen themselves. Explain why not.

## Activity 1.1: Investigating changes in pulse rate

How does your pulse rate change during exercise?

When you exercise, your muscle cells use more substances from the blood (for respiration). This means that exercise should cause the pulse rate (the number of heartbeats per minute) to increase. In this activity, you will plan an investigation to test this hypothesis.

You will need to think about variables, staying safe and the reliability of your data.

Something that can change in an investigation is a variable. You will change one **variable** (the **independent variable**) and measure changes in the pulse rate. The pulse rate is the **dependent variable**. You need to make sure that only the independent variable can affect the pulse rate. So you need to make sure that other variables (**control variables**) are kept the same.

A **hazard** is something that can cause harm. You will need to identify a hazard in your experiment and plan ways to control the **risks** (reduce the chances that the hazard will cause harm).

Generally, in an experiment you repeat your readings to check them. You can be more sure that your results are correct if repeated readings are similar. Repeated readings are more **reliable**.

**A1** Write down the hypothesis you will test.

**A2** Plan an investigation method to test the hypothesis. Make sure you include the variable you will change, the variable you will measure and any variables you will keep the same (control variables).

**A3** State at least *one* way to control risks.

**A4** Predict what will happen and explain your prediction.

**A5** Present your results neatly.

**A6** State how reliable you think your results are, and why you think this.

**A7** Compare your prediction with your results, and make a conclusion.

### Key terms

**control variable**: variable that you keep the same during an investigation.

**dependent variable**: variable you decide to measure in an experiment.

**hazard**: harm that something may cause.

**independent variable**: variable you decide to change in an experiment.

**reliable**: measurements are reliable when repeated measurements give results that are very similar.

**risk**: chance of a hazard causing harm.

**variable**: something you can measure or observe.

## Key facts:

✔ Red blood cells are used to carry oxygen.

✔ White blood cells destroy microorganisms that get into your body.

✔ Blood plasma carries the blood cells, and contains dissolved substances (such as nutrients and carbon dioxide).

## Check your skills progress:

- I can plan and carry out an investigation, and choose ways to change, measure and control different types of variable.
- I can plan ways to assess and control risk.
- I can compare results with a prediction in order to reach a conclusion.
- I can say whether my results are reliable enough to make a conclusion.

Chapter 1 . Topic 2

# Human respiratory system

You will learn:

- To describe the structure and function of the human respiratory system
- To describe how oxygen and carbon dioxide circulate between the blood and the air in the lungs
- To know that aerobic respiration occurs in animals and plants and releases energy in a controlled way
- To be able to use the summary word equation for aerobic respiration
- To plan investigations, including fair tests, while considering variables
- To choose experimental equipment and use it correctly
- To make predictions based on scientific knowledge and understanding
- To use results to describe the accuracy of predictions
- To take accurate measurements and explain why this matters
- To describe results
- To present and interpret scientific enquiries correctly
- To describe the application of science in society, industry and research
- To discuss the global environmental impact of science

## Starting point

| You should know that... | You should be able to... |
|---|---|
| Organs work together in organ systems | Plan to collect evidence in order to test a hypothesis |
| Animal and plant cells contain mitochondria, which are used for respiration | |

## Aerobic respiration

Blood carries oxygen and nutrients to all the cells in your body. One of these nutrients is glucose, which is made when large carbohydrates in your food are digested.

Your cells need glucose to give them the energy they need to carry out their functions and to make new substances. This process also needs oxygen from the air and so we call it **aerobic respiration**.

Aerobic respiration is a chemical reaction which occurs in the mitochondria of plant and animal cells.

Cells such as muscle cells, which require large amounts of energy, have lots of mitochondria.

In the process of aerobic respiration, oxygen and glucose react to produce carbon dioxide, water and energy. The energy produced is released to the cell in a slow and controlled way. We model the process using a **word equation**.

oxygen + glucose → carbon dioxide + water

The reactants are on the left of the arrow, and the products are on the right.

Your **respiratory system** (or 'breathing system'), your circulatory system and your digestive system all work together, to ensure that all the cells in your body get the substances they need for aerobic respiration.

**a)** What are the reactants in aerobic respiration?

**b)** Explain how these reactants get from outside the body to the cells.

### Key terms

**aerobic respiration**: respiration that requires oxygen to release energy from glucose.

**respiratory system**: group of organs that get oxygen into the blood and remove carbon dioxide. Also called the breathing system.

**word equation**: model showing what happens in a chemical reaction, with reactants on the left of an arrow and products on the right.

## Gas exchange and diffusion

Both plants and animals need oxygen for aerobic respiration. In humans, it is the function of **lungs** to allow oxygen into the blood. The lungs also allow carbon dioxide to leave the blood. This exchange of gases is called **gas exchange**.

Gas exchange also happens in plants. Their leaves have holes called **stomata**. Each **stoma** can be opened and closed by a pair of **guard cells**. When they are open, gases (such as oxygen, carbon dioxide and water vapour) can move into and out of the leaf.

Most gases in the air exist as tiny **particles** which are constantly moving. If there are many of one type of particle in one place and fewer in another, there is an overall movement of the particles towards the place with fewer particles. We call this **diffusion**. Gas exchange occurs when gases diffuse from one place to another.

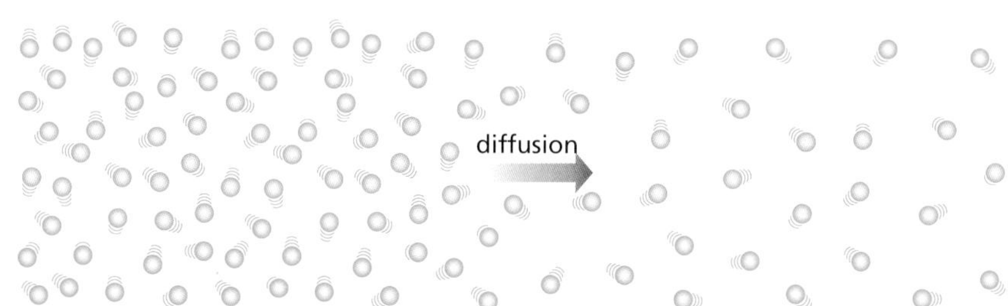


**1.6** *Diffusion of particles.*

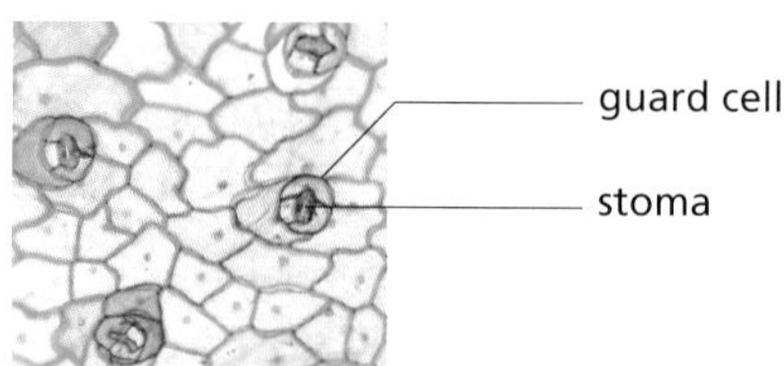


**1.5** *Stomata on a leaf. Magnification x300.*

### Key terms

**stoma** (plural stomata): hole in a leaf formed between two guard cells.

**guard cell**: cell that helps form a stoma in a leaf, to allow gases in and out.

**diffusion**: the spreading out of particles from where there are many (high concentration) to where there are fewer (lower concentration).

**gas exchange**: when two or more gases move from place to place in opposite directions.

## Gas exchange in the lungs

The **lungs** contain millions of pockets, called alveoli. Capillaries cover each **alveolus**. There are more oxygen particles inside an alveolus than in the blood in a capillary. This means that oxygen diffuses from the alveolus into the blood.

There are more carbon dioxide particles in the blood than in an alveolus. So, carbon dioxide diffuses from the blood into the alveolus.

### Key terms

**lungs**: organs that get oxygen into the blood and remove carbon dioxide.

**alveolus** (plural alveoli): tiny, pocket-shaped structure in lungs where gas exchange happens.

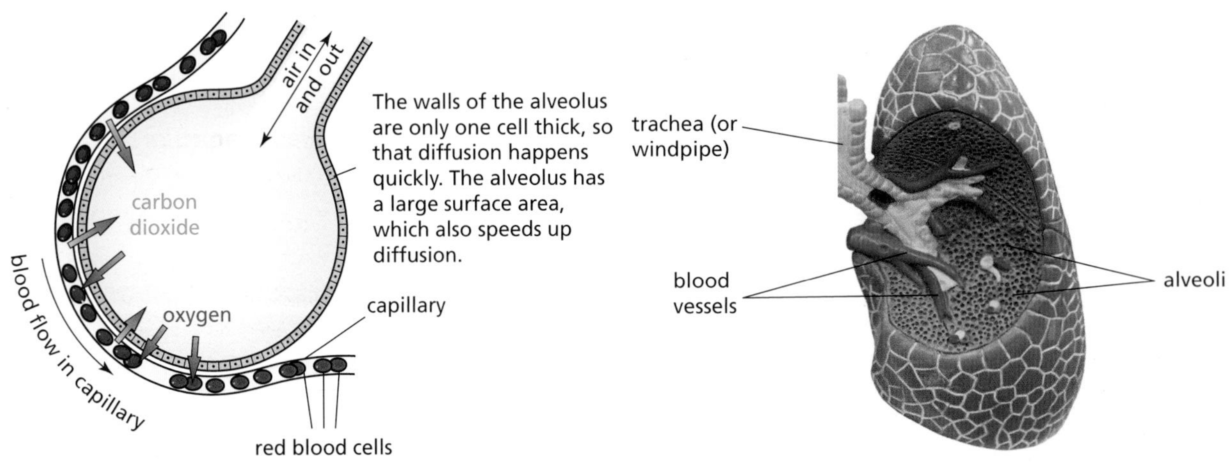


**1.7** *Gas exchange in an alveolus.*

**1.8** *A model of a human lung.*

2 Describe gas exchange in the lungs.

3 Explain why carbon dioxide particles move from the blood into an alveolus.

4 Look at figure 1.8. Explain why the lung looks like a sponge inside.

5 Why does an alveolus have a thin wall?

6 **a)** How do alveoli change the surface area of the inside of a lung?

**b)** Why is this adaptation important?

7 Explain how capillaries are adapted for gas exchange.

## Parts of the respiratory system

Your respiratory system moves air in and out of your lungs, through a series of tubes. The largest tube is the **trachea**, which is divided into two bronchi. Each **bronchus** goes into a lung, and leads to many smaller tubes called **bronchioles**.

Other parts of the respiratory system include the **ribs** and **diaphragm**.

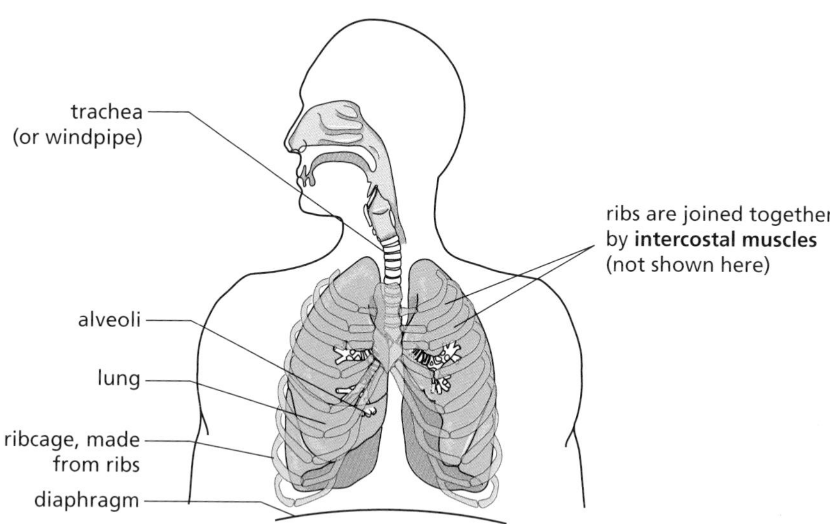


**1.9** *The parts of the respiratory system. The area inside your* ***ribcage*** *is your* ***chest****. Your lungs do not join to any part of your ribcage or your chest.*

8. Give another name for the windpipe.
9. The trachea divides into many smaller tubes. What structures are at the ends of all these tubes?

## Keeping it clean

Dust, viruses and bacteria may all damage your lungs. Your respiratory system has ways to stop them getting into your lungs.

Nose hairs filter air to remove larger particles. Smaller particles stick to a thick liquid **mucus**, produced by cells in the tubes of your respiratory system. These smaller particles include microorganisms, which can grow and reproduce in the mucus. However, **ciliated epithelial cells** use **cilia** to sweep the mucus up to your throat. You then swallow it and acid in your stomach kills the microorganisms.

### Key terms

**bronchioles**: small tubes leading from the bronchus in a lung.

**bronchus** (plural bronchi): large tube leading from the trachea into a lung.

**chest**: area inside the body between the ribcage, neck, backbone and diaphragm.

**diaphragm**: organ that helps with breathing.

**intercostal muscles**: muscles that join the ribs together, and move the ribcage to change the volume of your chest during breathing.

**rib**: bone that helps to protect your heart and lungs.

**ribcage**: all your ribs.

**trachea**: tube-shaped organ that allows air to flow in and out of your lungs.

**cilia**: waving strands that stick out of some cells.

**ciliated epithelial cell**: specialised cell with waving cilia to sweep mucus along.

**mucus**: sticky liquid that traps particles.

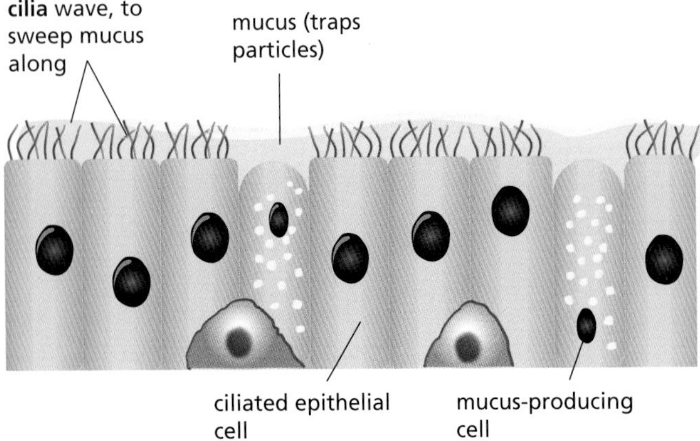


**1.10** *Ciliated epithelial cells help keep the lungs clean.*

**10** Why is it important to stop particles getting into your lungs?

**11** **a)** How are ciliated epithelial cells adapted for their function?

**b)** What other specialised cells are needed to keep the lungs clean?

## Muscles and breathing

The movement of muscles in the respiratory system is called **breathing**. When you **inhale** (breathe in), the intercostal muscles change shape. They get shorter and fatter, and we say that they **contract**. When they do this, they also move your ribcage upwards and outwards. Muscles in your diaphragm contract and flatten it. This increases the volume of your chest. Air flows through your nose into your lungs.

When you **exhale** (breathe out), the muscles relax. Your ribcage falls, and your diaphragm rises. Your chest volume gets smaller and air flows out of your lungs.

The number of times you inhale and exhale in one minute is your **breathing rate** (in breaths per minute).

### Key terms

**breathing**: movements of muscles in your respiratory system that cause air to move in and out of your lungs.

**breathing rate**: the number of times you inhale and exhale in one minute.

**contract** (muscle): when muscle tissue gets shorter and fatter it contracts.

**exhale**: breathing out.

**inhale**: breathing in.

**relax** (muscle): when muscle tissue stops contracting it relaxes.

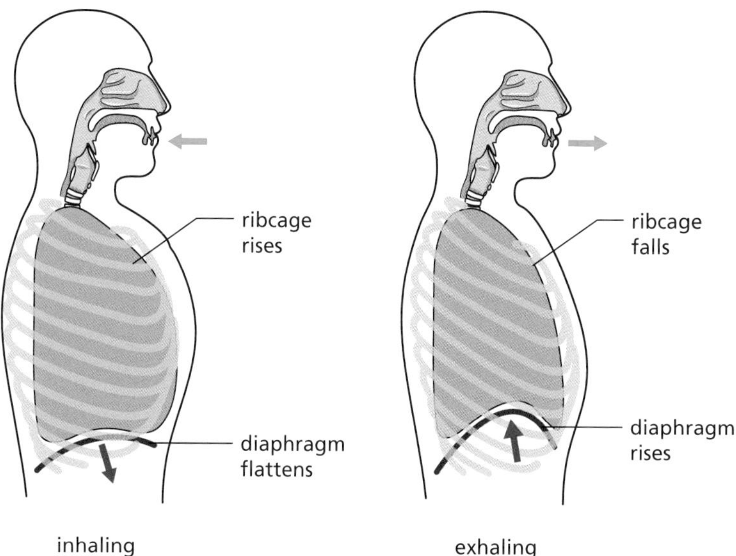


**1.11** *Inhaling and exhaling.*

**12** Place your hands on the front of your chest and inhale.

**a)** Describe what you feel.

**b)** Explain why you feel this.

**13** **a)** Describe how diaphragm muscles increase the volume of the chest.

**b)** What happens to your lungs when your chest volume increases?

**14** A student inhales and exhales six times in 30 seconds. What is the breathing rate?

### Key term

**precise:** how precise your measurement is depends on the measuring equipment and the smallest difference it can measure. The smaller the difference it can measure, the more precise the measuring equipment is.

## Activity 1.2: Investigating changes in breathing rate

What effect does exercise have on breathing rate?

When you exercise, your muscle cells use more substances from the blood (for respiration). In this activity, you will investigate whether exercise affects breathing rate.

Your results will be more reliable if they are repeatable (each repeated measurement has the same value). Repeated measurements that are close to one another are said to be **precise**. Repeated measurements that are not close to one another are less precise.

**A1** Write down the hypothesis you will test.

**A2** Plan an investigation method to test the hypothesis. Make sure you include the independent variable, the dependent variable and any control variables.

**A3** State at least *one* way to control risks.

**A4** Predict what will happen and explain your prediction.

**A5** Present your results neatly.

**A6** State how reliable you think your results are, and why you think this. Use the word 'precise' in your answer.

**A7** Compare your prediction with your results, and make a conclusion.

## Modelling the respiratory system

A model is a way of making something easier to understand. For example, figure 1.8 shows a physical model of a lung made out of plastic. A real lung is more complicated than this and the parts are more difficult to see.

The word equation for aerobic respiration is also a model and shows what happens in aerobic respiration, without a lot of the complicated detail.

We also use models to explain how things work. Figure 1.12 shows a model that explains how the movements of muscles in your respiratory system cause air to move in and out of your lungs.

In the model, the elastic membrane acts like our diaphragm. When the elastic membrane is pulled down, air is drawn into the jar through the glass tubing.

When the elastic membrane is released, it moves upwards back to its original position and air is pushed out of the jar through the glass tubing.

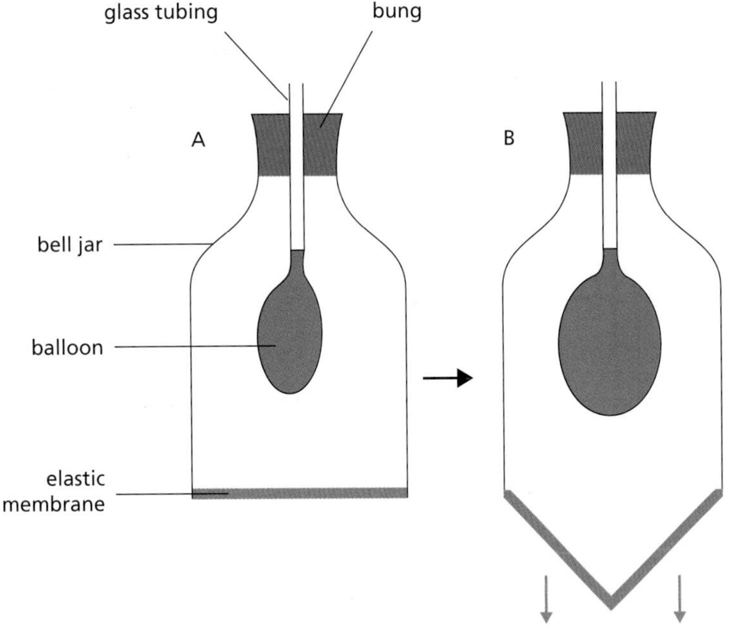


**1.12** *A model to explain how breathing causes air to move. The inside of the bell jar represents the inside of the chest.*

15 What is the difference between respiration and breathing?

16 Look at the model in figure 1.12.

**a)** What does the glass tubing represent?

**b)** What does the balloon represent?

**c)** What does the elastic membrane represent?

**d)** Imagine that you have pulled down the elastic membrane. Explain what happens to the air in the balloon when you release the membrane.

## Activity 1.3: Investigating lung volume

How can we measure lung volume?

We use the apparatus shown in figure 1.13 to measure 'lung volume'. This is the maximum volume of air you are able to exhale after a big breath. Adult men have an average lung volume of 4.8 litres. Adult women have an average lung volume of 3.1 litres.

Lung volumes in your class will show variation, and we can show how lung volumes vary on a bar chart using data that has been grouped together. To group your data, you will use a tally chart. An example (for hand measurements) is shown below.

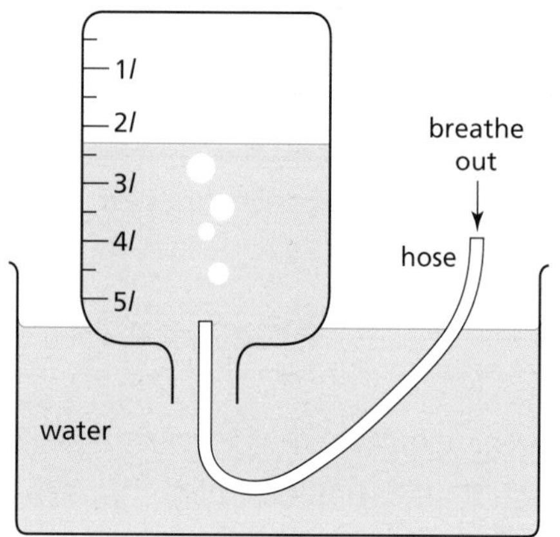


**1.13** *Apparatus to measure lung volume. When conducting this activity, always make sure to use a sterile mouthpiece, never share a mouthpiece with anyone else and take care not to breathe in.*

| Grouped widths of hands (mm) | Tally | Total |
|---|---|---|
| 80–84 | // | 2 |
| 85–89 | ~~////~~ | 5 |
| 90–94 | ~~////~~ /// | 8 |
| 95–99 | //// | 4 |
| 100–104 | / | 1 |

**Table 1.1** *Tally of different groups of hand widths.*

First divide your data into equal groups (e.g. lung volumes of 3.6–3.8 litres, 3.8–4.0 litres etc.). Then write the groups into your table. Then look at your lung volumes. One by one, cross out each value and put a mark in the 'tally column' to show its group. To complete your tally chart, add up each tally and write in the totals. You can then draw a bar chart using the totals from your tally chart. This type of grouped data is usually plotted on a bar chart without gaps between the bars.

**A1** Describe how to use the apparatus to measure lung volume.

**A2** State at least *one* way to control risks.

**A3** Predict how your lung volume compares with the values above. Explain your prediction.

**A4** Compare your prediction with your results.

**A5** Collect the lung volumes from everyone in your class and present your data as a tally chart.

**A6** Use your tally chart to draw a bar chart.

**A7** Evaluate your results. Remember, to do this think about what was good about the results and what could be improved. Think especially about reliability.

## Science in context: Air pollution

Vehicle engines that use petrol and diesel produce many substances that can harm our respiratory and circulatory systems. The World Health Organization thinks that about 7 million people die each year due to air pollution. Many cities are trying to reduce air pollution. In the United Kingdom, people are charged if they want to drive into the centre of London. In India, the authorities have banned all large diesel cars in Delhi. In Brazil, 70% of the people who live in Curitiba use very cheap buses to get around so that the streets are free from cars. Other ideas include banning cars from certain parts of cities, encouraging people to use bicycles and electric vehicles.

Think about what your country is doing or could do about air pollution. Choose two or three methods of reducing air pollution and evaluate them. (Write lists of advantages and disadvantages for each method and then decide which one you think would be the best to use.)

## Key facts:

✔ Breathing is the movement of the diaphragm and intercostal muscles, and it causes the lungs to inflate and deflate.

✔ The walls of the alveoli and blood vessels in the lungs are only one cell thick, to allow gas exchange to happen quickly by diffusion.

✔ Diffusion is the overall movement of a certain type of particle from where there are many of them to where there are fewer.

✔ Aerobic respiration occurs in mitochondria and requires glucose and oxygen, and produces carbon dioxide and water.

## Check your skills progress:

- I can carry out an investigation, comparing the results with a prediction in order to reach a conclusion.
- I can evaluate results in terms of reliability.
- I can use a tally chart to interpret data.

# Skeleton, joints and muscles

You will learn:

- To identify ball-and-socket and hinge joints
- To explain what makes bones move at hinge joints
- To describe the application of science in society, industry and research

## Starting point

| You should know that... | You should be able to... |
|---|---|
| Bones are organs that form part of the skeletal system | Follow a method carefully |
| | Record observations in a way that is clear and easy to read |

Aerobic respiration provides us with energy. One of the things that we need energy for is movement. Bones and muscles are organs that help us to move.

Which life process do bones and muscles help us with?

### Key terms

**joint**: place in your skeleton where bones meet.

**organ system**: group of organs working together.

**skeletal system**: group of organs that support the body and allow movement.

**skeleton**: another term for your skeletal system.

## The skeletal system

We are each born with 270 bones but adults only have 206 bones! This is because as we grow our bones grow too, and some of them join with others.

Which life process:

**a)** allows us to get from place to place

**b)** provides energy for us to get from place to place

**c)** causes the number of bones in our bodies to decrease as we get older?

An **organ system** is a group of organs that work together. Your bones work together and form your **skeletal system** or **skeleton**. This has three main functions:

- support – to hold parts of your body in certain positions
- protection – to stop parts of your body being damaged
- movement – to let you move (using **joints**).

The drawing below shows some parts of the human skeletal system and their jobs.

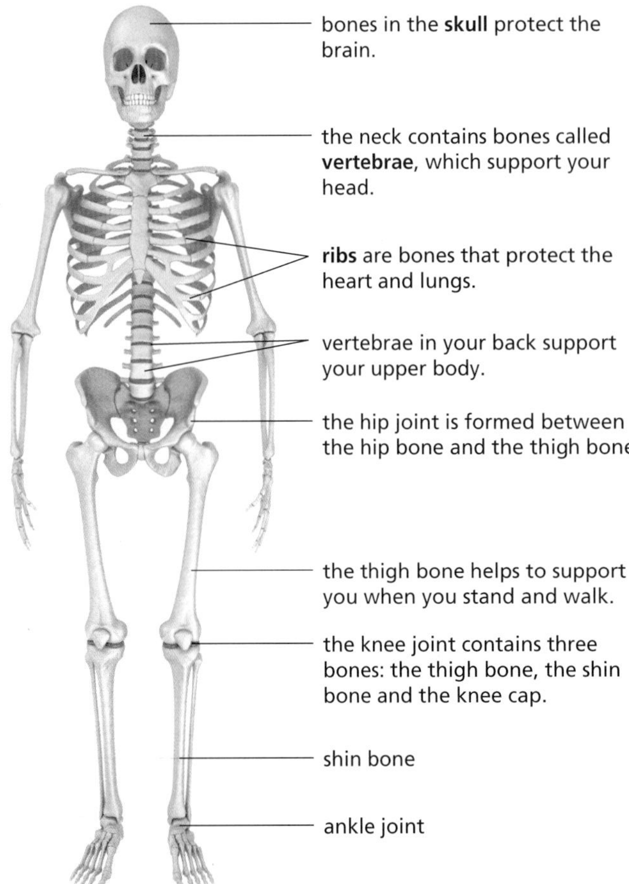


**1.14** *The human skeletal system.*

### Key terms

**rib**: bone that helps to protect your heart and lungs.

**skull**: a collection of bones that protect your brain.

**vertebra** (plural vertebrae): the bones in your back.

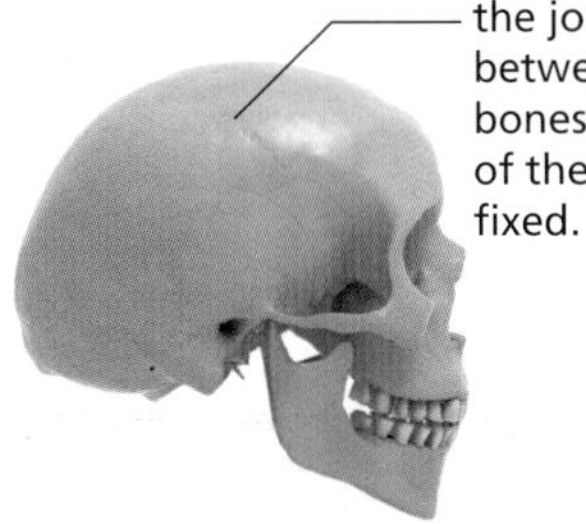


**1.15** *There are fixed joints in the skull.*

**3** a) What are the *three* main jobs of the skeletal system?

b) For each job, give the name of a bone that helps with this job.

**4** Give the names of the bones that form the knee joint.

## Joints

Bones meet at places called joints. The joints between many bones in the skull are fixed and cannot move.

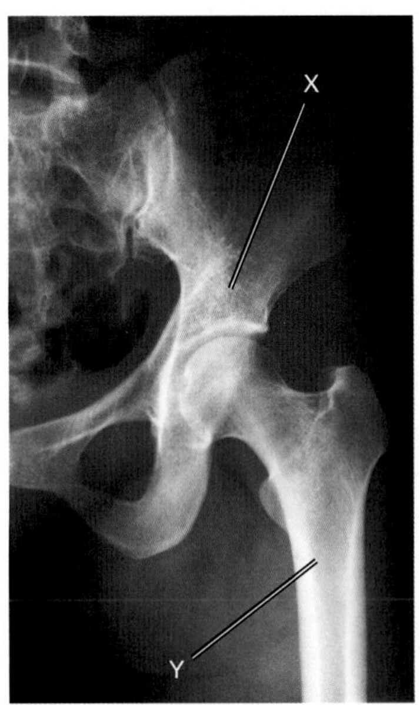


**1.16** *Doctors use X-rays to study the bones inside our bodies.*

Other joints allow movement. An example is the hip joint. Here, the top of the thigh bone forms a ball that fits into a socket in the hip bone. This type of joint is called a **ball and socket joint**. It allows movement in many different directions.

5 **a)** Look at figure 1.16. What are the bones labelled X and Y?

**b)** What *type* of joint is this?

The elbow is a **hinge joint** (shown in figure 1.17). Hinge joints allow movement in two opposite directions.

6 Your knee allows movement in two directions. What type of joint is it?

7 What type of joint is your shoulder? Give a reason for your answer.

## Muscles and joints

Ligaments are cords that hold the bones in a joint in position. Tendons attach the muscles to the bones.

Muscles move the bones. A muscle pulls on a bone when it contracts (get shorter and fatter). When a muscle is not contracted we say it is relaxed.

Muscles only pull and cannot push. So bones are often moved by **antagonistic pairs** of muscles. One muscle in a pair pulls a bone in one direction. The other muscle pulls the bone in the opposite direction.

Look at the drawings of bones and muscles in the arm in figure 1.17.

**a)** Describe the change in shape when a muscle contracts.

**b)** Name the muscle that contracts to raise the lower arm.

**c)** Describe what happens to both muscles when the arm is lowered.

**d)** Explain why there is a pair of muscles to move the lower arm up and down.

### Key terms

**ball and socket joint**: joint where a ball-shaped piece of bone fits into a socket made by other bones.

**hinge joint**: joint where two bones form a hinge.

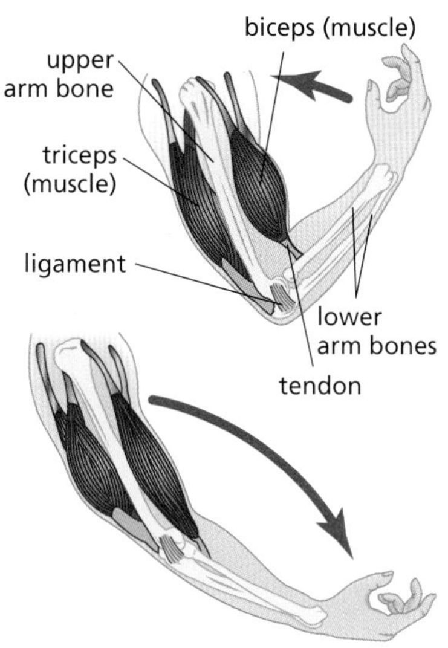


**1.17** *The bones and two of the muscles in the arm.*

### Key term

**antagonistic pair**: two muscles that pull a bone in opposite directions.

## Activity 1.4: Investigating arm muscles

Which arm muscles contract when you pull up or push down on a table edge?

Find a table (or a ledge) that will not move if you push down or pull up on it.

**A1** Read the instructions below and make predictions. Then do the activity.

- **A** Turn the palm of your *right* hand to face upwards.
- **B** Keep your hand flat, and put just your fingers under the edge of the table.
- **C** Use your *left* hand to grip the front of your *right* arm, at the top.
- **D** Pull your *right*-hand fingers upwards, as if lifting the table. Record what you feel.
- **E** Now, grip the back of your *right* arm, at the top.
- **F** Pull your *right*-hand fingers upwards again.
- **G** Put your *right*-hand fingers on top of the table (palm facing up).
- **H** Repeat steps C–F above but this time push your fingers downwards.

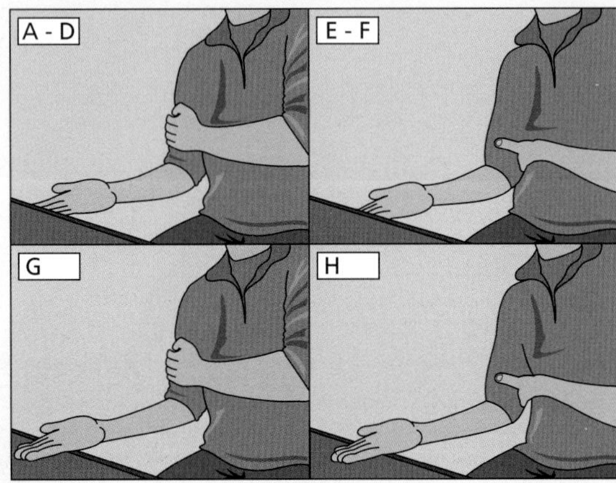


**1.18**

**A2** Record your observations.

**A3** Were your predictions correct?

**A4** Explain your observations using scientific knowledge.

## Science in context: Walking robots

Humans have tried to develop robots that can move like us for thousands of years. The earliest ones used clockwork springs as a source of energy, and later ones tried to use steam power. However, these robots could not operate on their own for very long, and they could not stand and walk on two legs in the same way that we can. This is because our muscles and bones, and the nervous system that controls them, is so complicated.

**1.19** *Scientists testing a robot with human-like movements.*

It is very difficult to do experiments on humans to find out why we are able to move in the ways we do. And the experiments that can be done involve so many variables that they cannot be fair tests. So, scientists study what happens when we move and then try to recreate these movements in robots.

Today, there are robots that can walk (and even run) in a human-like way. They are powered by batteries that last a long time and have complex circuits to control their balance and the movement of all the different parts. These robots can be used to explore places that are too dangerous for humans (such as volcanoes) and to help people with their daily lives (particularly those who have difficulty in moving).

## Injuries

Injuries may be painful and stop people moving easily. Common injuries include:

- fractures (broken bones)
- dislocations (when a bone comes out of place in a joint, which means that muscles cannot pull the bones in the correct directions)
- pulled muscles (when a muscle or a tendon has been stretched too much)
- sprains (when a ligament has been stretched too much).

**9** Look at the X-ray in figure 1.20.

**a)** Identify the joint that has been injured.

**b)** Give the name of this type of joint.

**c)** Give the name of this type of injury.

**d)** Give the reason why movement of the joint becomes very difficult with this type of injury.

**10** Give *one* similarity and *one* difference between a sprain and a pulled muscle.

**11** Do you think there are more antagonistic pairs of muscles in a ball and socket joint or in a hinge joint? Explain your answer.

**12** Sometimes the top of the thigh bone becomes worn and needs replacing. Explain *two* properties needed by a material used for a replacement.

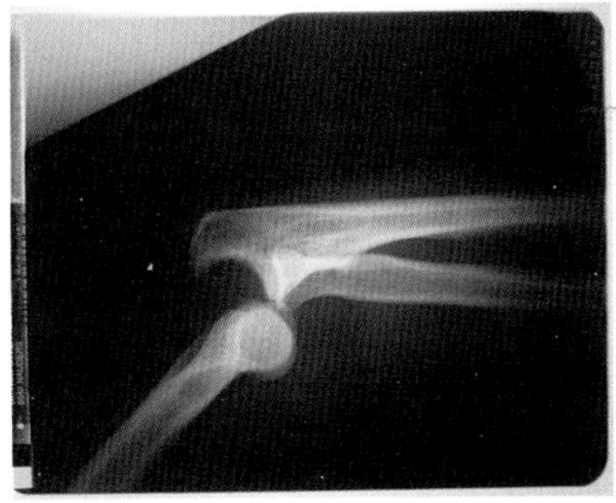

**1.20** *Injuries to the skeletal system may stop it working properly.*

## Activity 1.5: Investigating bones and joints

Copy and complete the following table. You may need to do some research to fill in all the gaps. The first row has been completed for you.

| Name of joint | Type of joint | Bones involved in the joint | Description of how the joint moves |
|---|---|---|---|
| elbow | hinge joint | humerus, radius and ulna | Joint moves in opposite directions. The biceps muscle contracts to pull the arm up and the triceps muscle contracts to pull the arm down. |
| shoulder | | | |
| hip | | | |
| knee | | | |

**Table 1.2** *Information about different types of joints.*

## Key facts:

- ✔ The skeletal system protects and supports parts of the body, and allows the body to move.
- ✔ Two important types of joint are ball and socket joints and hinge joints.
- ✔ Muscles cannot push and so antagonistic pairs of muscles are needed to move a bone in opposite directions.

## Check your skills progress:

- I can make predictions.
- I can explain observations using scientific language.

## End of chapter review

### Quick questions

1. Your heart, blood and blood vessels are all part of your:

   **a** respiratory system  **b** circulatory system

   **c** excretory system  **d** nervous system

2. The process by which ***plant*** cells release energy is called:

   **a** aerobic respiration  **b** photosynthesis

   **c** transpiration  **d** diffusion

3. A type of joint that allows movement in just two opposite directions is a:

   **a** sliding joint  **b** fixed joint

   **c** ball and socket joint  **d** hinge joint

4. When a muscle gets shorter and fatter it:

   **a** relaxes  **b** rejoices

   **c** contracts  **d** constricts

5. Another name for the trachea is:

   **a** bronchus  **b** windpipe  **c** oesophagus  **d** ventricle

6. Write out the word equation for aerobic respiration.

7. Describe the function of white blood cells

8. One function of the skeletal system is protection.

   **(a)** Give the name of an organ protected by the skull.

   **(b)** Give the name of an organ protected by the ribs.

   **(c)** State *two* other functions of the skeletal system.

## Connect your understanding

9. (a) Which parts of your respiratory system move to make your chest volume increase?

   (b) Explain why air flows into your lungs when your chest volume increases.

10. A student breathes in and out 7 times in 30 seconds. Calculate the breathing rate.

11. The drawing shows a part of a lung where gas exchange occurs.

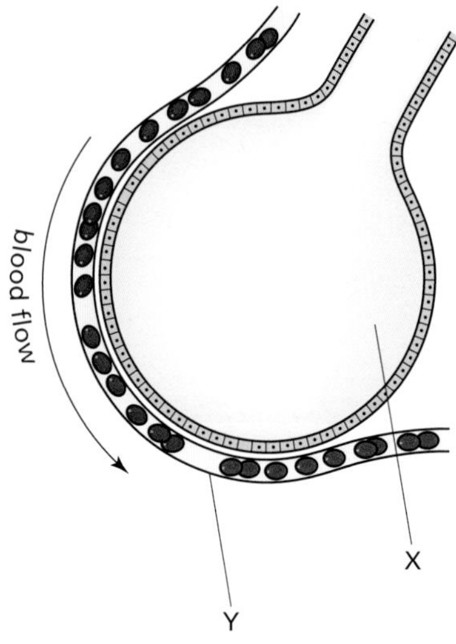


   (a) Give the names of parts X and Y.

   (b) Give the names of the gases that are exchanged.

   (c) What process causes the gases to move?

   (d) Explain how the structures shown in the diagram are adapted to help the gas exchange.

   (e) Explain why the red blood cells are not all the same colour in the diagram.

   (f) Explain *one* difference between the plasma at the top of the diagram and at the bottom of the diagram.

12. In an investigation, some athletes ran at different speeds for 5 minutes. A scientist then measured their pulse rates. The graph shows the results.

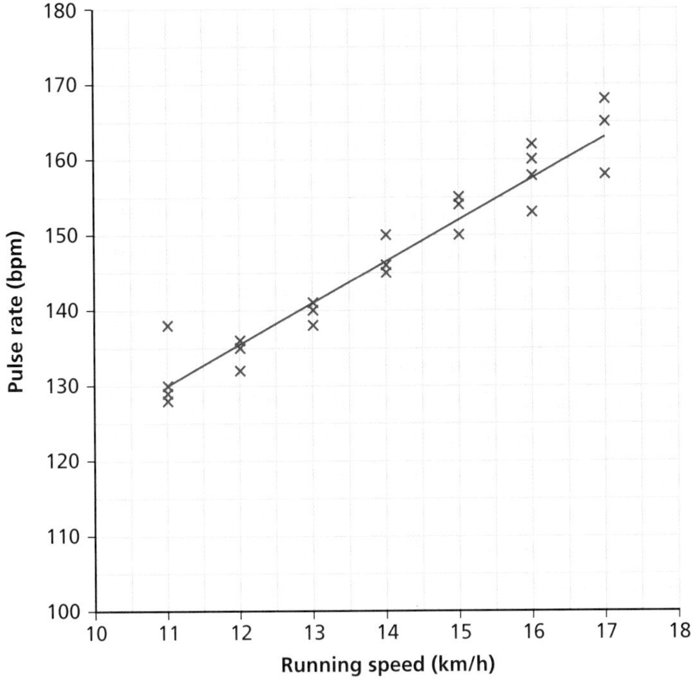


(a) What is the pulse rate a measure of?

(b) Use the graph to work out the pulse rate for an athlete running at 12.8 km/h.

(c) Describe the pattern of results on the graph. Start your answer using 'As the running speed increases, ...'

(d) Explain the pattern of results.

13. (a) Draw a labelled diagram to explain how specialised cells in the tubes of your lungs keep them clean.

(b) Explain what happens if these cells stop working.

**14.** The diagram shows some muscles and bones around the knee in a human leg.

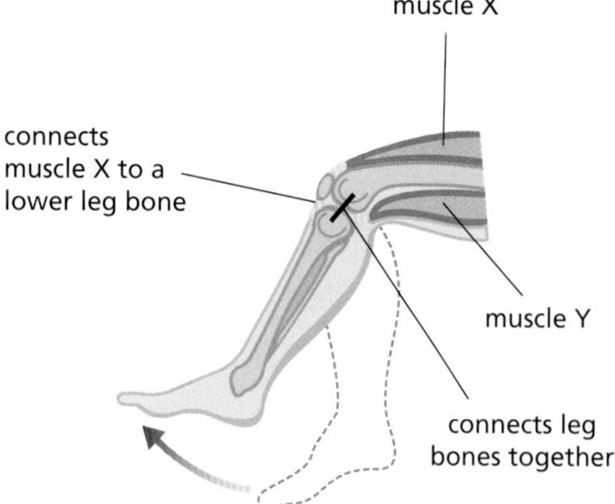


**(a)** Give the name of the part that connects a muscle to a bone.

**(b)** Give the name of the part that connects bones together.

**(c)** Which muscle has to contract so that the lower leg moves in the direction shown by the arrow?

**(d)** What happens to the other muscle during this movement?

**(e)** What are pairs of muscles like this called?

**(f)** Explain why bones must be moved by two muscles, rather than just one.

**(g)** Which life process do muscles help with?

**(h)** Give the name of the organ system formed by all the bones in the body.

**(i)** What type of joint is the knee joint?

## Challenge questions

**15.** Carbon monoxide is a poisonous gas, which can kill. It sticks to haemoglobin. This stops other substances sticking to haemoglobin. One of the early symptoms of carbon monoxide poisoning is an increase in breathing rate. Explain this symptom.

**16.** Write down *four* injuries that could occur at a joint, and for each one explain how the injury will affect movement.

2

# Chapter 2

## Nutrition

### What's it all about?

Each year, up to half a million people around the world become blind because they do not have enough vitamin A in their diets. In an effort to help, some scientists have made rice plants that produce 'golden rice' (shown in the upper half of this page). This rice contains a lot of vitamin A. It is hoped that people will grow this instead of normal rice. The scientists have used a technique called genetic modification to change the plants.

You will learn about:

- Balanced diets and why we need different types of nutrients
- What happens when we don't get enough of a nutrient or when we get too much of a nutrient
- How diet, drugs and fitness affect our bodies

You will build your skills in:

- Making and explaining predictions using scientific knowledge and understanding
- Identifying hazards and planning to control risks using risk assessments
- Identifying variables and planning to control some variables
- Using equipment correctly and taking accurate measurements

Chapter 2 . Topic 1

# A balanced diet

You will learn:

- To identify what makes a balanced diet for humans
- To describe the functions of the nutrients needed in a balanced diet
- That carbohydrates and fats can be used as an energy store in animals
- To identify and control risks in practical work
- To choose experimental equipment and use it correctly
- To take accurate measurements and explain why this matters
- To present and interpret scientific enquiries correctly
- To describe the application of science in society, industry and research

## Starting point

| You should know that... | You should be able to... |
|---|---|
| Animals and humans need nutrition | Recognise some dangers when doing experiments |
| Humans eat many different plants and animals | |

## Diets

Your **diet** is what you eat and drink. You need food for:

- energy
- growth and repair
- health.

A **nutrient** is any substance needed for energy or used as a raw material to make other substances. Your diet should contain the following nutrients:

- **proteins** (for growth and repair)
- **carbohydrates** (for energy)
- **fats** (also called **lipids**, and used to store energy)
- **vitamins** and **minerals** (for health, growth and repair)
- **water** (for dissolving and carrying substances around the body and for temperature control through sweating).

**Fibre** is not classed as a nutrient because it is not digested or absorbed by the body. You do need fibre in your diet

### Key terms

**carbohydrate**: nutrient needed for energy. Examples include starch and sugars (such as glucose).

**diet**: what you normally eat and drink.

**fats**: nutrients needed by your body to store energy.

**lipids**: another word for fats.

**minerals**: nutrients that living organisms need in small amounts for health, growth and repair. Also called mineral salts.

though to keep your intestines (gut) healthy. Fibre forms most of your solid waste or **faeces**. A lack of fibre may cause **constipation**, when your intestines become blocked. Good sources of fibre include wholemeal bread, brown or wholegrain rice, cereals, lentils, nuts and fruits.

1. List *three* nutrients that your body needs.
2. a) Why is fibre not a nutrient?
   b) Why does your body need fibre?
   c) List *two* good sources of fibre in your diet.
3. Give an example of a carbohydrate.

### Key terms

**nutrient**: a substance that an organism needs to stay healthy and survive.

**proteins**: nutrients you need for growth and repair.

**vitamins**: nutrients you need in small amounts for health, growth and repair.

## Food labels

It is the law in most countries that foods should have labels on them to show the amounts of nutrients and fibre they contain. These labels can help people to make healthier choices about what they eat.

Food labels also show the energy in foods. Your body needs energy for growing, moving and keeping warm.

We measure energy in units called **joules (J)**. There are 1000 joules in 1 kilojoule (kJ). You need between 8000 and 10 000 kJ each day. You need more energy if you are more active and growing quickly.

| NUTRITION INFORMATION | | |
|---|---|---|
| | Per 40 g serving | Per 100 g |
| Energy | 600 kJ (143 kcal) | 1500 kJ (358 kcal) |
| Protein | 4.7 g | 11.8 g |
| Carbohydrate | 20.9 g | 52.3 g |
| *sugars* | *10.0 g* | *25.0 g* |
| *starch* | *10.9 g* | *27.3 g* |
| Fat | 1.9 g | 4.7 g |
| Calcium | 200 mg | 500 mg |
| Iron | 27 mg | 68 mg |
| Vitamin C | 10 mg | 25 mg |
| Fibre | 9.9 g | 24.8 g |

We also measure energy in calories (cal) or kilocalories (kcal). There are 1000 cal in 1 kcal.

Many food labels show the amounts of different types of carbohydrate.

We measure nutrients and fibre in grams (g) or milligrams (mg). There are 1000 mg in 1 g.

**2.2** *Food labels show what is in 100 g of a food and what a normal serving of the food contains. To compare foods you must use the amounts per 100 g.*

**2.1** *These foods are good sources of fibre.*

### Key terms

**constipation**: when your intestines become blocked.

**faeces**: solid waste material produced by humans and other animals.

**fibre**: food substance that cannot be digested but which helps to keep your intestines healthy.

**joule**: unit used to measure energy.

4 Look at the food label in figure 2.2.

a) How many grams of protein are in 100 g of the food?

b) How large is one serving?

c) How many grams of fat are in one serving?

d) Give the name of *one* mineral in the food.

e) How many milligrams of calcium would be in 80 g of this food?

f) How much energy is in 100 g of the food?

g) Why does your body need energy?

h) What process does your body use to release energy from food?

5 Give the name of a nutrient that is a good source of energy.

6 Explain why more active people need to eat more food.

7 a) What is 500 mg in grams (g)?

b) What is 4.7 g in milligrams (mg)?

c) What is 25 mg in grams?

8 Look at the food label in figure 2.2. If you only ate this food, about how much would you need to provide your energy for one day?

## Risk assessments

A **hazard** is the harm that something may cause. The chance of a hazard causing harm is a **risk**.

When scientists are using a chemical substance in an experiment, they need to do a **risk assessment** to make sure they stay safe. There are three steps:

- identify the hazards
- identify who could be harmed by the hazards
- plan ways to control the risk of harm for everyone.

### Key terms

**hazard**: harm that something may cause.

**risk**: chance of a hazard causing harm.

**risk assessment**: the process of identifying the hazards involved in a practical investigation and deciding how to control them.

| Substance | Hazards | Controlling the risks |
|---|---|---|
| iodine solution | stains skin and clothing | use a dropper to gently add the solution and avoid splashes |
| | | keep clothes away from iodine solution |
| | stings if it gets in your eyes | wear eye protection |
| biuret solution | corrosive – attacks skin | use a dropper to gently add the solution and avoid splashes |
| | corrosive – can cause eye damage | wear eye protection |

**Table 2.1** *Risk assessment information for food testing (note that food tests are not covered in Stage 8).*

## Carbohydrates

Carbohydrates are a group of compounds made from carbon, hydrogen and oxygen. Some carbohydrates are small particles and are soluble in water. These are **sugars**. Examples of sugars include sucrose (the sugar that you use at home) and glucose. Some carbohydrates are large, insoluble particles. An example is starch.

All carbohydrates are excellent stores of energy, which is why they are important for humans. Respiration releases energy from the carbohydrates in your diet.

Starch is the main carbohydrate that we eat. There are many good sources of starch, including rice, pasta, potatoes and bread.

Sugars are carbohydrates found in sweet foods, such as candy and cakes. However, sugary foods may damage teeth.

If your body does not use all the carbohydrates you eat, it turns them into fats and stores them. Overeating carbohydrates makes people get larger.

**9** What does your body need carbohydrates for?

**10** Name a source of starch in your diet.

**11** Give *two* reasons why people should not eat lots of sweet things.

### Key term

**sugar**: soluble carbohydrate, which exists as small particles. Glucose is an example.

**2.3** *These foods are good sources of starch.*

**2.4** *You should only have small amounts of sugary foods.*

## Activity 2.1: Investigating energy in food

How do we compare the energy released by different foods?

Burning releases energy from foods quicker than respiration. Figure 2.5 shows how to heat cold water using burning food. The more energy released, the more the water temperature rises.

burning food on a mounted needle
boiling tube
water
clamp and stand

**2.5** *Burning food samples to measure temperature rise.*

Here are the instructions for this experiment.

**A** Add 20 $cm^3$ of cold water to a boiling tube.

**B** Measure the temperature of the water.

**C** Use a balance to find the mass of the food.

**D** Heat the food until it starts to burn.

**E** Heat the water using the burning food, until it stops burning.

**F** Measure the temperature of the water again.

**G** Calculate the change in the temperature using this equation:

change in temperature = temperature at end – temperature at start

We use this experiment to find the different amounts of energy in foods. The variable we change (the **independent variable**) is the food. The variable we measure (the **dependent variable**) is the rise in temperature.

To make a fair comparison between foods, we must keep other variables the same. These are **control variables**, and include the volume of water and the mass of food.

The results of some experiments are in the table.

| Food | Temperature of water at start (°C) | Temperature of water at end (°C) | Rise in temperature (°C) |
|---|---|---|---|
| bread | 17.2 | 22.6 | |
| cookie | 18.5 | 28.5 | |
| popcorn | 17.9 | 25.8 | |

**Table 2.2** *Rise in temperature of water when being heated by burning food.*

**A1** Write a risk assessment for this experiment.

**A2** Copy the table and calculate the rise in temperature for each food.

**A3** Which food contained the most energy?

**A4** What was used to measure the water temperature?

**A5** Apart from water volume and mass of food, suggest *one* more control variable.

**12** Look at figure 2.6. What are the temperatures on thermometers A, B and C?

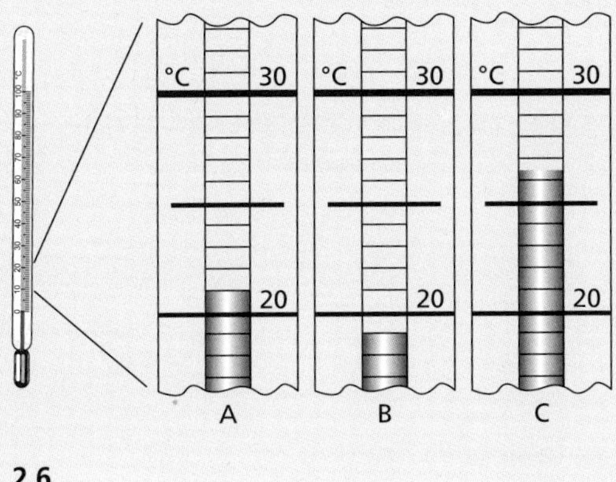


**2.6**

## Key terms

**control variable**: variable that you keep the same during an investigation.

**dependent variable**: variable you decide to measure in an experiment.

**independent variable**: variable you decide to change in an experiment.

## Fats

Your body uses fats (lipids) to store energy. Fat is also stored under your skin to help keep you warm. Good sources of fats include butter, cheese, eggs, oils and milk.

Overeating foods containing fats makes people larger.

**13** Name a source of fats in your diet.

**2.7** *Some good sources of fats.*

## Proteins

Your body uses proteins to make substances to build new cells, repair your body and to let you grow. Muscles are mainly protein and so meats are a good source. Other good sources of protein include beans, cheese, eggs, fish, lentils, milk and tofu.

People who need lots of protein in their diets include athletes and body builders, and young people who are growing quickly.

**14** Name a source of protein in your diet.

**15** Give *one* source of protein that is not from an animal.

**16** Explain why an athlete needs to eat lots of protein.

**2.8** *Some good sources of proteins.*

## Vitamins and minerals

You need small quantities of vitamins and minerals in your diet. Each has a function that keeps your body healthy and working properly. Table 2.3 gives some examples.

| Vitamin or mineral | Good sources | What it does |
|---|---|---|
| vitamin A | carrots, cheese, eggs, oily fish | keeps eyes and skin healthy |
| vitamin C | lemons, limes, oranges, pineapples | keeps skin and gums healthy |
| vitamin D | oily fish, red meat, eggs | keeps bones and teeth strong |
| calcium (a mineral) | bok choy, milk, cheese | used in bones and teeth |
| iron (a mineral) | beans, meat, dark green vegetables | used in red blood cells |

**Table 2.3** *Some examples of vitamins and minerals.*

**17** **a)** Name a vitamin and a source of that vitamin in your diet.

**b)** What does the vitamin do?

**18** **a)** Name a mineral and a source of that mineral in your diet.

**b)** What does the mineral do?

## Balanced diets

No single food contains all the nutrients you need and so you should eat foods from many sources. If you eat many different foods and get nutrients in the correct amounts, you have a **balanced diet**.

### Key term

**balanced diet**: eating many different foods to get the correct amounts of nutrients.

**2.9** *For a balanced diet, you should eat lots of fruits and vegetables. You also need foods containing starch. Foods containing lots of fats and sugars should form only a small part of your diet.*

## Science in context: Recommendations for a balanced diet

To help people eat a balanced diet, governments often make recommendations. These recommendations have different names in different countries. Table 2.4 shows some 'Reference Intakes' from the European Union. These amounts are only a guide for adults. People of different ages and with different levels of activity will need to eat slightly more or slightly less than these amounts.

| | EU Reference Intake |
|---|---|
| energy | 8400 kJ (2000 kcal) |
| protein | 50 g |
| carbohydrate | 260 g |
| sugars | 90 g |
| fat | 70 g |

**Table 2.4** *Reference Intakes published by the European Union.*

**19** Table 2.4 recommends a maximum of 90 g of sugars in a day.

**a)** Name the type of carbohydrate that you should eat more of than sugars.

**b)** Why should less of the carbohydrate in your diet come from sugars?

**20** How does the recommended energy in the table compare to the amount *you* need? Explain your answer.

## Activity 2.2: Investigating food recommendations

What are the food recommendations in different countries?

**A1** Use different books and/or the internet to discover how much of each nutrient your country recommends. Also discover the values from a country far from yours.

**A2** Present your information as a table, to compare the values.

## Key facts:

- ✔ Your diet needs to contain nutrients (carbohydrates, fats, proteins, minerals and vitamins) for energy, growth and repair, and health.
- ✔ A balanced diet contains the right amount of the different nutrients from a wide variety of different foods.
- ✔ You also need water and fibre in your diet.
- ✔ Carbohydrates and fats store energy, which is measured in joules (J) or kilojoules (kJ).

## Check your skills progress:

- I can identify independent, dependent and control variables.
- I can write a risk assessment.

### Making links

In Stage 7 you may have learned about energy transfers. Describe the main energy transfers in your body.

# The effects of lifestyle on health

You will learn:

- To discuss the effects of lifestyle on humans' development and health
- To evaluate secondary information for relevance and potential bias
- To present and interpret measurements and observations in a suitable way
- To describe results in terms of any trends and patterns and identify any abnormal results
- To reach conclusions studying results and explain their limitations
- To describe the application of science in society, industry and research

## Starting point

| You should know that... | You should be able to... |
|---|---|
| A healthy diet contains many different nutrients | Accurately plot graphs |
| Smoking can cause diseases | |

They way in which we lead our lives is called our lifestyle. Someone's lifestyle can affect their growth, their development and their health. Some people have more choices about their lifestyle than others.

## The effects of diet

People who do not receive enough of a certain nutrient may have a nutritional deficiency. This will affect their health, growth and development.

As well as obvious nutritional deficiencies, there is evidence to suggest that the lack of some nutrients has other effects. The graph in figure 2.10 shows how boys' height may be affected by the amount of protein they eat. The data is from over 150 countries.

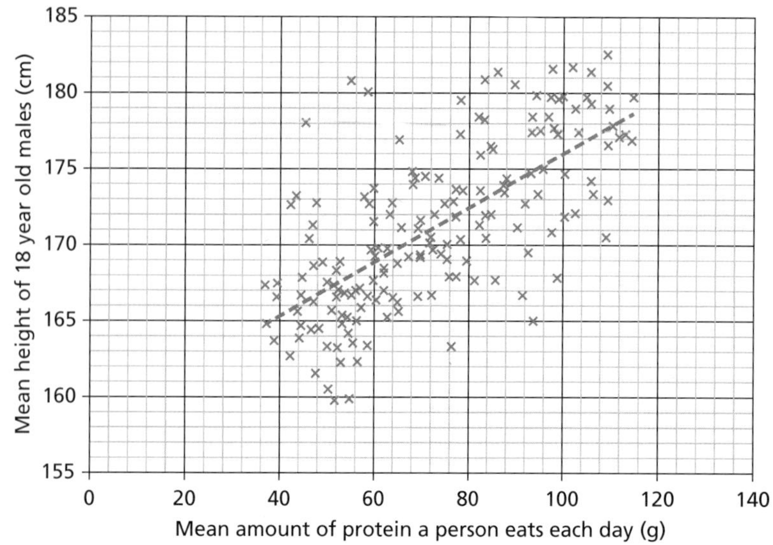

**2.10** *How height may be affected by the amount of protein eaten.*

You can see a **trend** on the graph. Graphs are used to look for links between two variables. The graph shows that as the amount of protein increases, the height also increases. The link does not prove that eating more protein causes an increase in height. However, it suggests to scientists that this may happen and that further studies should be done.

### Key term

**trend:** pattern seen in data in which there is a change in a certain direction.

1. Explain why scientists are doing more studies on how the amount of protein in the diet affects growth.
2. Explain how a lack of calcium in the diet affects growth and development.
3. Suggest why people who do not eat enough protein are less strong than they could be.

**2.11** *Obesity is caused by overeating and lack of exercise.*

## Obesity

As well as a lack of nutrients, an unhealthy diet can also have too much of a particular nutrient. An example is **obesity**.

If your body does not use all the carbohydrates you eat, it turns them into fats and stores them. Overeating carbohydrates and fats makes people larger. A lack of exercise makes the problem worse. If a person gains a lot of weight, they may become obese, which means that their health is in danger because of their weight.

Health problems caused by obesity include **type 2 diabetes** and **high blood pressure**.

In type 2 diabetes, too much glucose is dissolved in your blood plasma. This can damage organs, such as the heart and eyes.

### Key terms

**obesity**: being so overweight that your health is in danger.

**type 2 diabetes**: disease in which there is too much glucose in your blood, which can damage organs.

If the pressure of your blood in your blood vessels is too high, it can cause them to burst. This can also cause damage to organs such as the heart and brain. High blood pressure can also be caused by eating too much salt.

Eating too much fat is thought to cause some blood vessels to get blocked, particularly those going to the heart muscle. This can cause **heart disease**, in which the heart does not pump very well.

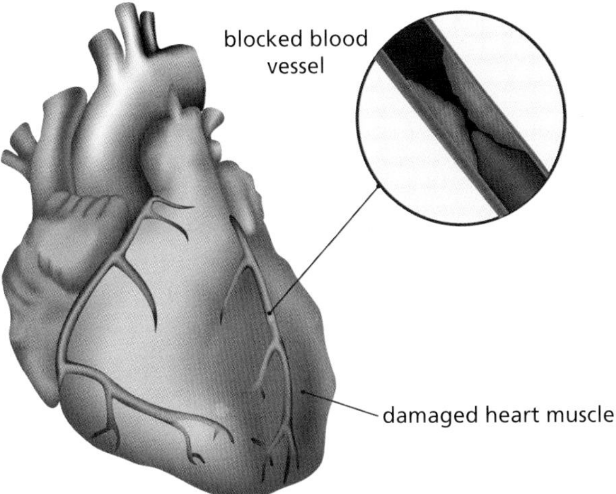


**2.12** *Heart disease.*

4. What factor in the diet can cause high blood pressure?
5. List ***four*** possible effects of having too much fat in the diet.
6. Explain why a lack of exercise can cause obesity.
7. A stroke is when part of the brain is damaged. It can be caused when blood stops getting to part of the brain. Explain two ways in which this might occur.

## Fitness and exercise

Fit people easily do everything they need to do each day (such as running upstairs and not being out of breath). Like having a healthy, balanced diet, being fit will also keep you healthy. Many people make lifestyle choices to stay fit by taking regular exercise (such as walking or playing sports). Exercise strengthens your muscles and bones that help you move. It also strengthens your heart muscles making the heart better at pumping your blood.

### Key terms

**heart disease**: when the heart does not work well.

**high blood pressure**: when the pressure of blood inside your blood vessels is at risk of bursting them.

## Drugs

A drug is a chemical that changes the way your body works. Drugs can be used for many different purposes. Some drugs help to fight disease, some reduce pain and others are decongestants (drugs that can provide relief from a blocked nose). Drugs that are used for medical purposes are **pharmaceutical drugs**. Other drugs are used for enjoyment and many of these recreational drugs are **illegal** in many countries.

**2.13** *Another name for a medicine is a pharmaceutical drug.*

Because drugs affect how your body works, they may also change how it grows and develops.

Give *three* different uses for pharmaceutical drugs.

### Key terms

**illegal drug**: drug that individual people are not allowed to buy or use. Different countries have different laws about drugs.

**pharmaceutical drug**: drug used in healthcare to help the body fight a disease, or to relieve pain.

## Smoking

Many people smoke tobacco (mainly in cigarettes). Tobacco smoke can damage cells in your respiratory system. This includes the ciliated epithelial cells in the tubes leading into your lungs, which help to keep your lungs clean.

The heat and substances in smoke damage the cilia, and paralyse them. They no longer wave and so cannot sweep up mucus. Smoking for many years destroys the cilia. Long-term smokers must cough up the mucus made by the mucus-producing cells to get rid of it. This is a 'smoker's cough'.

In the lungs of a smoker, why do the cilia stop working?

10 Explain why people who smoke often have to cough at regular intervals.

11 Explain why long-term smokers are more likely to get lung infections.

12 Coughing, and the substances in smoke, can destroy alveoli. This causes people to feel breathless because of poor gas exchange. Explain why destruction of alveoli causes poor gas exchange.

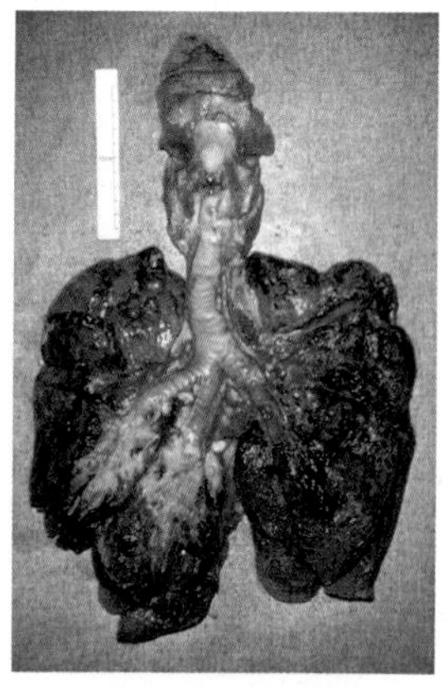

**2.14** *Lungs showing the effects of tobacco smoke.*

Tobacco smoke contains a drug called **nicotine**, which affects your brain and increases heart rate. This drug is **addictive**, which means that people feel that they need to have it. Over time, nicotine damages blood vessels and can make them more likely to become blocked.

**13** a) Describe one instant effect that nicotine has on your body.

b) Explain how nicotine damages the circulatory system with time.

Tobacco smoke contains a black, sticky mixture of substances called **tar**. This may cause **cancer**, in which a tissue makes new cells in an uncontrollable way. The new cells may form a lump called a **tumour**.

**14** What is a cancer tumour?

**15** Look at figure 2.14. Lung tissue is pink. Why is the tissue not pink in this lung?

Scientists ask questions about why things happen. They then create hypotheses that might explain why these things happen. They test those hypotheses using experiments or observations.

Between 1900 and 1950 in the United Kingdom, there was a big increase in lung cancer. A doctor, Richard Doll, created three possible hypotheses to explain this. He thought that the number of people with lung cancer may depend on an:

- increase in road building (and the tar used)
- increase in traffic (and exhaust fumes)
- increase in smoking.

He made many observations. He found that the more people smoked, the more lung cancer deaths there were. The two variables changed together. This is a link.

We use a **line graph** to look for a link between two variables. These variables must both be measured as numbers. You plot the variables on a graph and look for a trend. For example, you might look for a line of points. You might be able to draw a **line of best fit** through the points.

### Key terms

**addictive**: substance that makes people feel that they must have it.

**cancer**: when cells in a tissue start to make many copies of themselves very quickly.

**nicotine**: addictive drug in tobacco smoke.

**tar**: sticky black liquid found in cigarette smoke.

**tumour**: a lump of cancer cells.

### Key terms

**line graph:** graph of two variables, both measured in numbers.

**line of best fit**: straight or curved line drawn through the middle of a set of points to show the pattern of data points.

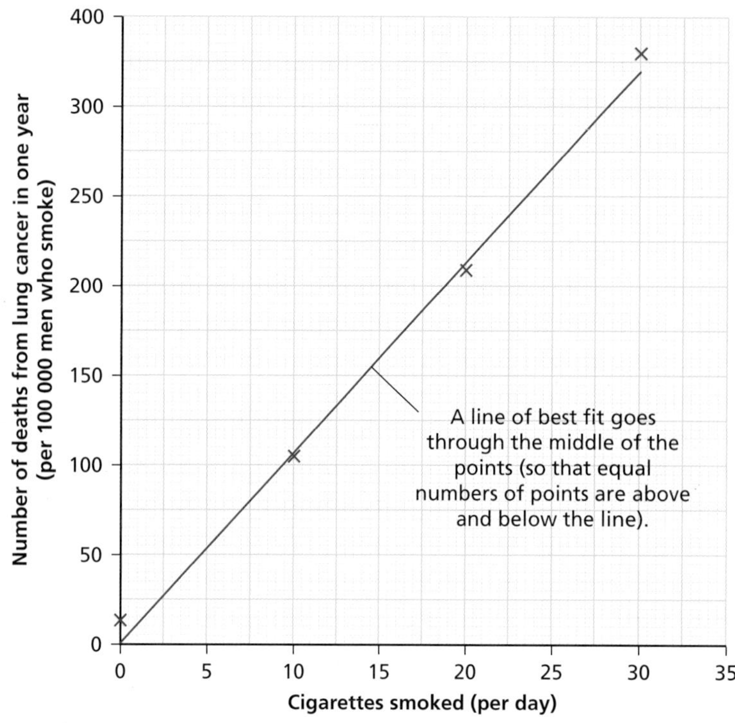


**2.15** *The line on this graph shows a link between the number of cigarettes smoked and the number of deaths from lung cancer.*

## Science in context: Smoking tobacco

A scientific study found that between 1990 and 2015, smoking caused one in every 10 deaths in the world. It also found that the number of smokers in the world is increasing, because the world population is increasing.

The percentage of people who smoke in some countries has fallen (such as Brazil, Australia and Nigeria). In others, there was no change (such as Bangladesh, Indonesia and the Philippines). The percentage of smokers in some groups has increased (such as women in Russia).

Smoking creates jobs, such as tobacco farming. Governments also get money from taxes on tobacco. However, smoking causes diseases and many countries try to reduce the number of smokers. They put warnings on cigarettes and raise the taxes on them.

## Activity 2.3: Investigating smoking

What are the 'pros' and 'cons' of smoking?

**A1** Use this and other books and/or the internet to:

- find some good points about smoking
- find some bad points about smoking
- discover what your country is doing about smoking.

**A2** Write a report of three or four paragraphs. End your report by saying what you think about smoking. Your teacher may ask you to use your ideas in a debate.

**16** Scientists use the apparatus in figure 2.16 to investigate cigarette smoke.

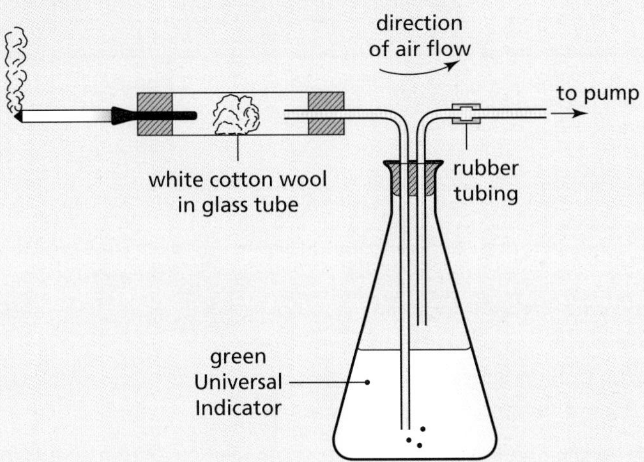


**2.16**

**a)** Predict what happens to the colour of the cotton wool. Give a reason for your prediction.

**b)** A scientist puts a thermometer in the glass tube. Will the temperature reading be higher, lower or the same as outside the apparatus?

**17** Tar coats the inside of the lungs. Explain what effect this has on gas exchange.

## Key facts:

- ✔ Eating too much food that is rich in fat or carbohydrate can cause obesity.
- ✔ Obesity can cause problems with the heart and blood vessels, so blood does not flow so well.
- ✔ Tobacco smoke paralyses cilia, and contains an addictive drug called nicotine.
- ✔ Tar in tobacco smoke causes cancer.

## Check your skills progress:

- I can look for trends and patterns in data.
- I can interpret line graphs (and lines of best fit).

### Making links

In Stage 7 Chapter 4 Topic 2 you learned about indicators. The Universal Indicator in the apparatus in figure 2.22 turns orange. What does this tell you about cigarette smoke?

In Stage 7 Chapter 5 Topic 1 you learned about combustion. If the Universal Indicator is replaced by limewater, describe what will happen.

## End of chapter review

### Quick questions

1. In your body, carbohydrates that you do not need for energy are converted into:

   a fats  b proteins

   c sucrose  d glucose

2. An important mineral for bones is:

   a sodium  b iron

   c calcium  d vitamin D

3. Another name for the trachea is the:

   a bronchus  b windpipe

   c oesophagus  d air sac

4. A poor diet can lead to obesity.

   **(a)** What is obesity?

   **(b)** Describe the type of diet that causes obesity.

5. **(a)** Give *one* reason why we need to eat each of these nutrients:

   - carbohydrates
   - fats
   - proteins.

   **(b)** Name *one* other nutrient that we need.

   **(c)** Describe what may happen if we do not eat enough of that nutrient.

6. Name the addictive drug in tobacco smoke.

## Connect your understanding

7. A scientist uses some burning foods to heat up some water. She wants to know which food contains the most energy.

| Food | Temperature of water at start (°C) | Temperature of water at end (°C) |
|---|---|---|
| dried potato | 20.1 | 23.0 |
| cookie | 20.1 | 24.8 |
| popcorn | 20.1 | 24.2 |

   **(a)** Which variable (shown in the table) did she control?

   **(b)** Suggest one other control variable.

   **(c)** Which food contained the most energy?

   **(d)** Explain your answer to part **c**.

   **(e)** What units is energy measured in?

8. When Noor is 40, a blockage starts to form in one of his blood vessels. It grows as he gets older.

   **(a)** Suggest a reason why the blockage formed.

   **(b)** Explain why the blockage causes Noor's blood pressure to rise.

   **(c)** Describe one problem that high blood pressure can have.

   **(d)** At age 65, a doctor tells Noor that he has heart disease. Explain how this has happened.

## Challenge questions

9. The table shows some information about different foods in a meal.

| Values per serving | Beans | Avocado | Rice |
|---|---|---|---|
| Energy | 500 kJ | 700 kJ | 850 kJ |
| Protein | 16 g | 2 g | 4.6 g |
| Carbohydrate (including sugars) | 4 g (4 g) | 0.8 g (0.7 g) | 48 g (0.3 g) |
| Fat | 2 g | 29 g | 3.6 g |
| Calcium | 20 mg | 0.024 g | 2.5 mg |
| Iron | – | 1 mg | 4 mg |
| Fibre | – | 7 g | 0.6 g |

(a) A scientist says "This meal is balanced".
Suggest what she means.

(b) A student says "This shows that rice contains more energy than beans or avocados".
Explain why the student cannot say this.

(c) Calculate the number of milligrams of calcium in this serving of avocado.

**10.** In an experiment, a scientist gave people different amounts of nicotine.
The scientist then measured the speed of the blood flowing through capillaries in the skin.
After the experiment, the speed of the blood quickly returned to normal.

| Mass of nicotine taken into the body (mg) | Average speed of blood in the capillaries ($mm^3$/sec) |
|---|---|
| 0 | 0.000 047 |
| 0.50 | 0.000 036 |
| 0.98 | 0.000 034 |
| 1.90 | 0.000 027 |

(a) What is the link shown in this data?

(b) Suggest what nicotine does to capillaries to cause this effect.

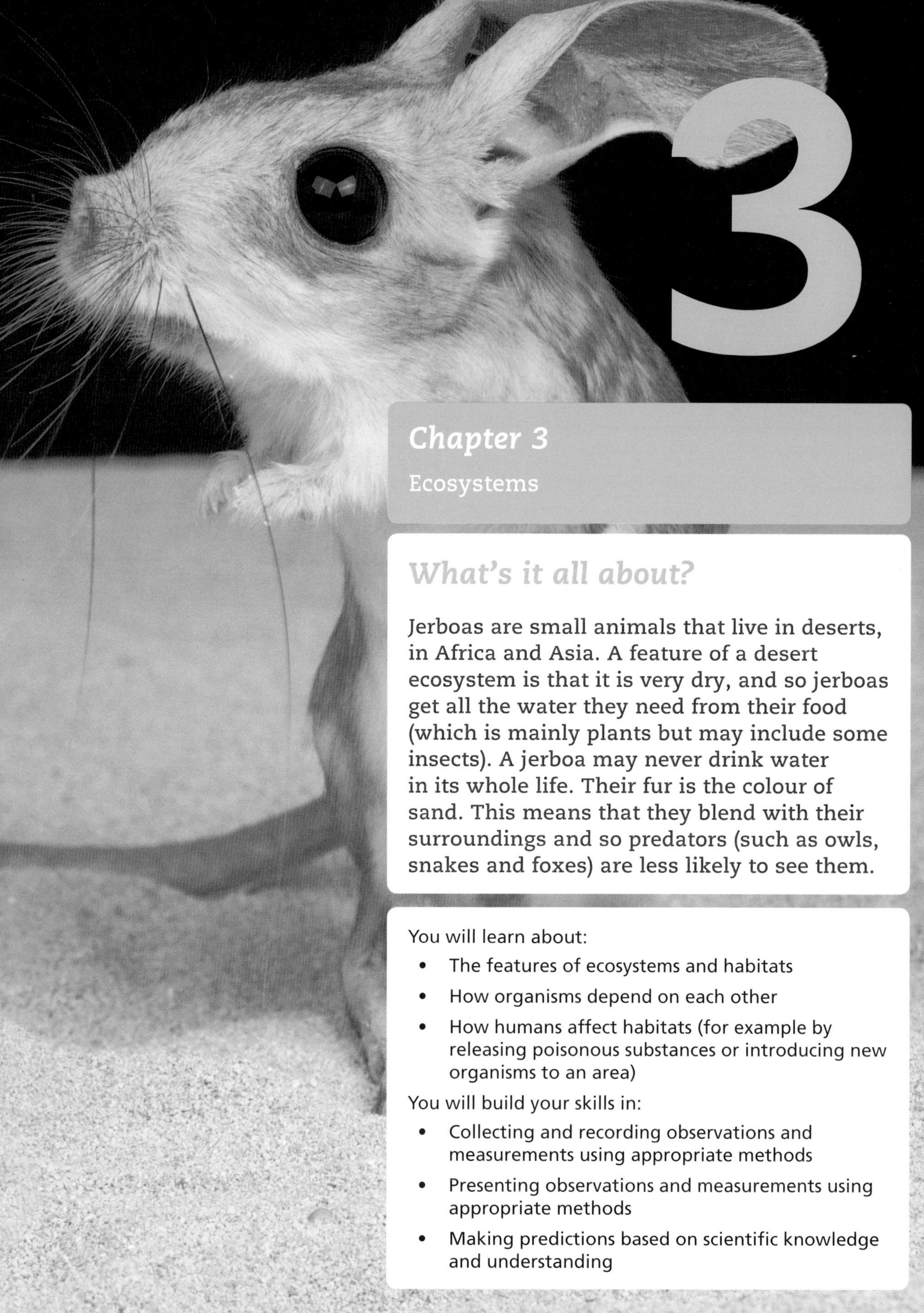


# Chapter 3

Ecosystems

## What's it all about?

Jerboas are small animals that live in deserts, in Africa and Asia. A feature of a desert ecosystem is that it is very dry, and so jerboas get all the water they need from their food (which is mainly plants but may include some insects). A jerboa may never drink water in its whole life. Their fur is the colour of sand. This means that they blend with their surroundings and so predators (such as owls, snakes and foxes) are less likely to see them.

You will learn about:

- The features of ecosystems and habitats
- How organisms depend on each other
- How humans affect habitats (for example by releasing poisonous substances or introducing new organisms to an area)

You will build your skills in:

- Collecting and recording observations and measurements using appropriate methods
- Presenting observations and measurements using appropriate methods
- Making predictions based on scientific knowledge and understanding

Chapter 3 . Topic 1

# Habitats and ecosystems

You will learn:

- To identify different ecosystems on Earth
- To recognise the range of habitats in an ecosystem
- To carry out practical work safely
- To collect and record observations and measurements appropriately
- To sort organisms through testing and observation
- To present and interpret scientific enquiries correctly
- To evaluate secondary information for relevance and potential bias
- To choose experimental equipment and use it correctly
- To discuss how scientific knowledge is developed
- To describe the application of science in society, industry and research
- To discuss the global environmental impact of science

## Starting point

| You should know that... | You should be able to... |
|---|---|
| Different types of organisms are classified into different species | Present data using tables |
| A habitat is the place where an organism naturally lives | Draw a bar chart |
| | Make careful observations and measurements |

## Habitats

The place where a certain species of organism naturally lives is its **habitat**. Some habitats are large, such as sand dunes in a desert. Some habitats are small, such as a pond or the bark of a tree.

The things that organisms need are **resources**. A habitat provides an organism with all the resources it needs, such as:

- water
- shelter and protection
- food.

An organism's habitat contains non-living parts, such as temperature, light, wind, water and rocks. These are **physical factors**. It may also contain living factors, that is, other organisms such as trees.

### Key terms

**habitat**: the place where an organism lives.

**physical factor**: non-living part of an environment (e.g. wind).

**resource**: anything that is needed or used by an organism.

The surroundings of an organism are its **environment**. An environment contains:

- other organisms
- a range of physical factors.

## Ecosystems

All the organisms in a certain area depend on one another for things, such as food and shelter. All the organisms that depend on one another in an area, together with all the physical factors in that area, form an **ecosystem**.

Examples of ecosystems include coral reefs, deserts and rainforests. There are often many habitats within one ecosystem. For example, in a rainforest there are some organisms that live up in the trees. This is a different habitat to the habitat on the ground underneath the trees, and these two habitats are different to the habitat in which small insects live in the bark of a tree. The bark is the outer layer of the tree trunk.

### Key terms

**ecosystem**: all the organisms and the physical factors in an area.

**environment**: the other organisms and physical factors around a certain organism.

**3.1** *This savanna ecosystem is in Kenya. A savanna is an area of open grassland with some trees.*

**3.2** *There are many different habitats in a coral reef ecosystem.*

**a)** What is a habitat?

**b)** Name *two* habitats.

**c)** Name *one* habitat in which fish live.

2 a) Name the ecosystems in figures 3.1 and 3.2.

b) List the different organisms living in figure 3.1.

c) Look at the giraffes in figure 3.1. Describe their environment.

d) Describe *two* different habitats in this savanna ecosystem.

3 Describe your environment now.

4 a) Describe *two* physical factors in a rainforest ecosystem.

b) Describe how these factors would be different in a habitat at the top of the trees compared with a habitat on the ground.

5 Fungi are **decomposers**. Most fungi live in dark, warm and wet conditions. Suggest a habitat in which fungi might live.

### Key term

**decomposer**: microorganism that causes decay.

## Activity 3.1: Investigating choice of habitat by woodlice

**3.3** *Woodlice.*

Woodlice live in many ecosystems, including forest ecosystems. What type of habitat do woodlice prefer? In this investigation, you are going to give woodlice a choice of four different environments and see where they choose to go. The environments can be light and damp, light and dry, dark and damp, or dark and dry.

You are using live creatures in this investigation and you must show care and respect for them. Always handle them with a plastic spoon and paintbrush to ensure that you do not damage them. Always wash your hands thoroughly after handling woodlice, and always return the woodlice back to where they were found.

Figure 3.4 shows the apparatus to use – it is called a choice chamber.

**A1** Write a step-by-step method describing how you will use the choice chamber to test which environment woodlice prefer.

**A2** What observations and results will you collect?

**A3** How will you ensure that your results are reliable?

**A4** Draw a results table and record your results.

**A5** Draw a chart or graph to show your results. Decide which type of chart or graph is best for your results.

**A6** Describe what your chart or graph shows.

**A7** Write a conclusion for your investigation. State what you have discovered and explain why you think this. You may need to do some research about the natural habitat of woodlice to help you.

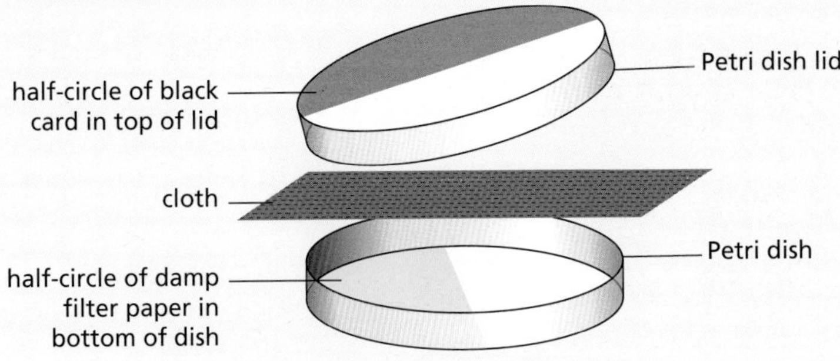


**3.4** *A simple choice chamber.*

## Sampling organisms in a habitat

To discover what lives in an ecosystem, scientists examine small parts of it. These small parts are **samples**. The photos show some ways of collecting samples.

**3.5** *A **pitfall trap** is a container buried in the ground, with a cover to stop rain getting in. It traps animals that fall into it. Pitfall traps have different sizes.*

**3.6** *A **quadrat** is a square frame. Scientists place it in an area and look for the different organisms inside it.*

### Key terms

**pitfall trap**: jar buried in the ground to collect small animals that walk on the ground.

**quadrat**: square frame used to take samples in a habitat.

**sample**: small portion of something, used to discover what the whole of the thing is like.

*3.7 Scientists often collect samples using nets. This 'mist net' is used to catch small birds (which are later released). Other nets are used in water and long grass.*

To examine small animals, scientists need to handle them carefully and look at them closely. The photos below show some ways to do this.

**3.8** *If you carefully suck on one tube of a **pooter**, small animals are sucked into the collecting jar.*

**3.9** *To examine and count small, crawling animals without harming them, we use soft brushes to move them.*

**3.10** *We magnify small animals and parts of plants with a **magnifying glass** (or **hand lens**).*

By studying organisms using methods like these, scientists find out what lives in the different habitats in an ecosystem. They can also work out the number (or **frequency**) of each species in an ecosystem. The number of a certain species living in a certain ecosystem is called its **population**.

### Key terms

**frequency**: the number of times something occurs.

**hand lens**: another term for magnifying glass.

**magnifying glass**: used to make things appear bigger (magnify them).

**pooter**: device to suck small animals into a collecting jar without harming them.

**population**: the total number of individual organisms of one species living in a certain area.

6 Why do scientists use soft brushes to handle small organisms?

7 What type of organism does a pitfall trap collect?

8 Why do we use a magnifying glass to examine small animals?

9 Describe a type of animal that you could find out about using a quadrat.

## Science in context: Protecting species

It is not just scientists who look for organisms and count their numbers. All around the world there are organisations that encourage everybody to report the organisms they see around their homes. People then upload the information about what they have seen using the internet or an app.

**3.11** *Stag beetles are rare in some parts of Europe.*

For example, in some parts of Europe stag beetles are becoming very rare. They are no longer found in Denmark or Latvia, and scientists are worried that the species could die out altogether (become extinct). Wildlife organisations in Europe ask people to report any stag beetles they find. This helps scientists to work out where the beetles can still be found and how common they are in those places. This knowledge will help the scientists to plan ways to save the species.

## Presenting results

Scientists took samples of a savanna ecosystem in East Africa to discover which grasses lived there. In the results table, the ticks show the grasses found in each sample.

| Type of grass | Sample number | | | | | | | | | | | |
|---|---|---|---|---|---|---|---|---|---|---|---|---|
| | 1 | 2 | 3 | 4 | 5 | 6 | 7 | 8 | 9 | 10 | 11 | 12 |
| elephant grass | ✓ | ✓ | ✓ | | | ✓ | | ✓ | | ✓ | | |
| pan dropseed | | | | | | | | | | | | |
| red dropseed | | ✓ | | ✓ | ✓ | | | | | | | |
| red grass | | ✓ | | ✓ | | | ✓ | | ✓ | ✓ | ✓ | ✓ |

**Table 3.1** *Grasses found in a savanna ecosystem.*

A bar chart makes it easier to compare frequency data.

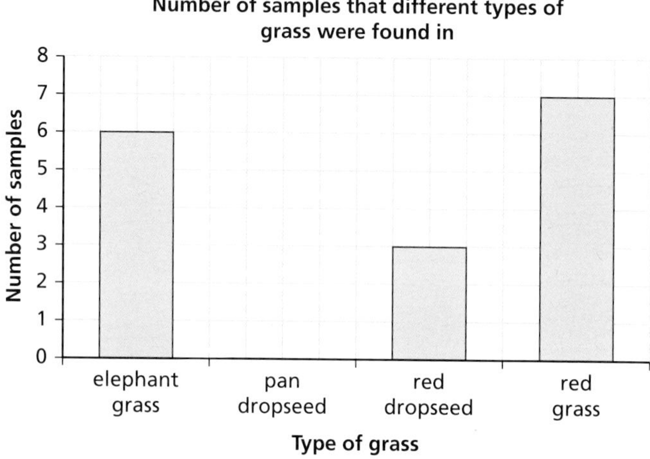


**3.12** *Bar chart showing the number of samples that different grasses were found in.*

**10** Look at the table and the bar chart.

**a)** Which was the most common grass in this area?

**b)** Which grass was not found in this area?

**c)** Suggest which method was used to collect these samples.

**d)** Why is it better to show the results in a bar chart rather than in a table?

**e)** Why would a line graph not be an appropriate way to present these results?

**11** In part of the Amazon rainforest in Brazil, some students left pitfall traps for one night. The table shows the number of species of beetle in each trap the next morning. Present these results using a bar chart.

| Type of beetle | Pitfall trap number | | | | | | | | | |
|---|---|---|---|---|---|---|---|---|---|---|
| | 1 | 2 | 3 | 4 | 5 | 6 | 7 | 8 | 9 | 10 |
| C *Coprophanaeus* | | | | ✓ | | | | | ✓ | |
| D *Dichotomius* | ✓ | | ✓ | ✓ | ✓ | ✓ | | ✓ | ✓ | ✓ |
| E *Eurysternus* | | | | | | | | | | |
| O *Onthophagus* | | ✓ | | ✓ | | | ✓ | | ✓ | ✓ |

**Table 3.2** *Number of beetles caught in pitfall traps.*

**12** Look back at figure 3.1.

**a)** Draw a table to show the number of each different animal in the photo. Only include those animals that you can clearly see.

**b)** Present your results as a bar chart.

**c)** A scientist says that the giraffe is the most common animal in this ecosystem. Does the evidence in your table and bar chart support this conclusion?

**13** A study of a desert found 800 jerboas, 10 000 desert grass plants, 15 tawny eagles and 70 sand cobras. Plot a bar chart to show the populations of these organisms.

14 Using a quadrat, a scientist takes samples of a field. The total area of the samples taken is 20 $m^2$, in which the scientist finds 104 dandelion plants. The total area of the field is 5000 $m^2$. Calculate the total number of dandelion plants you would expect to be in the whole field.

## Activity 3.2: Investigating your local habitats

What types of organisms live in a habitat near you?

**A1** Choose a habitat near you and describe it. You could include a drawing.

**A2** Make a prediction about what organisms live there.

**A3** Decide how you will take samples from your habitat.

**A4** Write a risk assessment for your investigation. Write down the hazards you might find and how they may cause harm to people. Then explain how you will control the risks of harm from these hazards.

**A5** Carry out your investigation. You may need to use books (field guides) to help you identify some of the organisms that you find.

**A6** Present your results as a table (and as a bar chart, if you can).

**A7** Make a conclusion about which organisms are the most and least common.

## Key facts:

- ✔ An ecosystem is a set of living things that depend on one another and a set of physical environmental factors found in a particular area.
- ✔ Examples of ecosystems include tropical rainforest, desert, grassland and coral reef.
- ✔ A habitat is the place in an ecosystem where a certain organism can find the resources it needs.
- ✔ An organism's environment is the physical factors and the other organisms in its surroundings.

## Check your skills progress:

- I can choose and use a variety of different sampling techniques to study organisms in their habitats.
- I can plan to collect reliable data.
- I can identify hazards and write a risk assessment.
- I can use frequency data to construct a bar chart.
- I can use data to make conclusions.

### Making links

In Stage 7, Chapter 3 you learned about classification. Suggest how scientists can find the names of the organisms in the samples they collect using nets or pitfall traps.

# Bioaccumulation in food chains

You will learn:

- To describe the effect of toxic substances in an ecosystem
- To describe the application of science in society, industry and research
- To discuss the global environmental impact of science

## Starting point

| You should know that... | You should be able to... |
| --- | --- |
| Food chains and food webs show how energy passes from one organism to another in the feeding relationships found in an ecosystem | Model energy flow in an ecosystem using food chains and food webs |

Organisms need food for energy. They release the energy using respiration.

Plants produce their own food. They are **producers.** Animals need to consume (eat) other organisms. They are **consumers.**

Animals that only eat plants are **herbivores.** Those that only eat other animals are **carnivores.** Those that eat plants and animals are **omnivores.**

1. Which life process do organisms use to release energy?
2. Is a tree a producer or a consumer? Give a reason for your choice.
3. **a)** Name a herbivore in your country.
   **b)** Name a carnivore in your country.
   **c)** Use *two* of the **bold** words above to describe yourself.

A **food chain** is a model that helps us to understand how energy passes through the organisms in an ecosystem. Each stage of a food chain shows a different trophic level.

Producers are in the first **trophic level.** When they are eaten, energy passes to the next trophic level – the **primary consumers.** When the primary consumers are eaten, energy passes from them to the **secondary consumers.**

### Key terms

**carnivore**: animal that eats other animals.

**consumer**: animal that eats other living things.

**food chain**: diagram showing feeding relationships in a habitat – each species is a food source for the species at the next level up.

**herbivore**: animal that eats plants.

**omnivore**: animal that eats both plants and animals.

**primary consumer**: animal that eats plants (producers). These are the second trophic level in an ecosystem. For example, an antelope eats grass.

**producer**: organism that makes its own food, such as a plant.

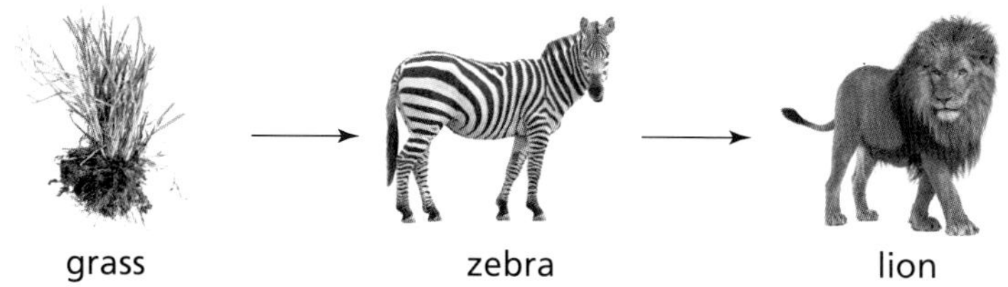


**3.13** *A food chain from an African savanna.*

**Key terms**

**secondary consumer**: animal that eats a primary consumer. These are the third trophic level in an ecosystem. For example, a hyena eats an antelope.

**trophic level**: level in an ecosystem. All producers are in the first trophic level. Energy passes from the lower to the higher trophic levels.

**4** Look at figure 3.13 and identify:

**a)** a producer

**b)** a consumer

**c)** a carnivore

**d)** a herbivore

**e)** an organism at the second trophic level.

**5** In a rainforest in Borneo, small mammals called grey tree rats feed on fig trees. Borneo pythons (a type of snake) eat grey tree rats.

**a)** Write a food chain for these organisms. (Do not draw pictures.)

**b)** Underneath the names of the organisms, write one or more of these words:

| | |
|---|---|
| carnivore | first trophic level |
| herbivore | primary consumer |
| producer | secondary consumer |
| second trophic level | third trophic level |

**6** **a)** Jerboas in the Gobi Desert eat plant leaves, seeds and insects. Which of these words best describes a jerboa?

carnivore herbivore omnivore

**b)** Why did you choose this word?

**c)** Give *one* way in which food chains are a good model for understanding energy flow in an ecosystem, and one way in which they are not a good model.

The arrows in a food chain show energy flow. Energy from the Sun is trapped by the producers in the first trophic level. When the primary consumers eat the producers, energy flows from the first trophic level to the second trophic level.

When the secondary consumers eat the primary consumers, energy flows from the second trophic level to the third trophic level. However, at each step, some of the energy is lost. Not all the energy in one trophic level passes to the next. Figure 3.14 shows some of the ways in which energy loss happens.

**3.14** *Energy loss in a food chain.*

**7** a) Where do plants get their energy from?

b) Where does a secondary consumer get its energy from?

c) The secondary consumers get less energy from the primary consumers than the primary consumers got from the producers. Explain why.

**8** The third consumer in a food chain is the tertiary consumer. The fourth one is the quaternary consumer.

In a grassland ecosystem, aphids feed on grass. Sparrowhawks hunt smaller birds, such as thrushes. Aphids are prey for spiders. Thrushes are predators of spiders.

a) Write a food chain for these organisms.

b) Label the organisms in your chain with words to describe their nutrition.

## Bioaccumulation

Dead organisms and wastes are broken down by decomposers. However, some substances are not broken down and can be absorbed by plants and microorganisms.

If an animal eats an organism that contains these substances and the substances are not excreted, then the substances will build up in its cells and tissues. This is called **bioaccumulation**.

### Key term

**bioaccumulation**: build-up of a substance in an organism because the substance cannot be broken down and is not excreted.

If an organism contains too much of a substance, it may be harmed. For example, mercury is a poisonous metal used in gold mining and is often released from the mines. The bioaccumulation of mercury in some birds can make it difficult for them to fly. Mercury also affects the nervous system of humans, making people twitch and have difficulty walking. It can kill.

**9** **a)** Describe *one* effect of mercury bioaccumulation in birds.

**b)** Suggest a reason why this effect may cause a bird to die.

Organisms in higher trophic levels are more likely to be harmed by substances such as mercury than organisms at lower levels. This is because the substance builds up through the food chain.

In figure 3.15 the red dots represent mercury. The algae absorb mercury from the water. Shrimp are small creatures that live in the water and eat algae. When the shrimp eat the algae, they get the mercury. The fish eat many shrimp and so they get all their mercury. The fishing birds eat lots of fish and get all the mercury from those fish. The fishing birds contain so much mercury that they die.

**3.15** *A model to represent the bioaccumulation of mercury.*

**10** Look at figure 3.15.

Explain why the fishing birds get ali the mercury that was in the algae.

**11** In the 1950s, a plastics factory was putting mercury into the sea in Minamata Bay (in Japan). Suggest why 70 people who lived along the shore of the bay died.

**12** A substance called dichlorodiphenyltrichloroethane, commonly called DDT, used to be used by farmers in the 1950s and 1960s to kill insects. If there is enough DDT in a bird, it causes them to lay eggs with very weak shells. Explain why the population of peregrine falcons decreased at this time, even though the population of blackbirds (which they eat) was unaffected. Blackbirds eat earthworms, which can absorb DDT from the soil.

**3.16** *A peregrine falcon*

## Science in context: Using knowledge about bioaccumulation

Scientists use their knowledge about bioaccumulation to provide advice. They regularly test fish to see how much mercury they contain, and organisations use this data to advise people on how much of different sorts of fish they can eat. An example of an advisory leaflet from a community in California is shown below.

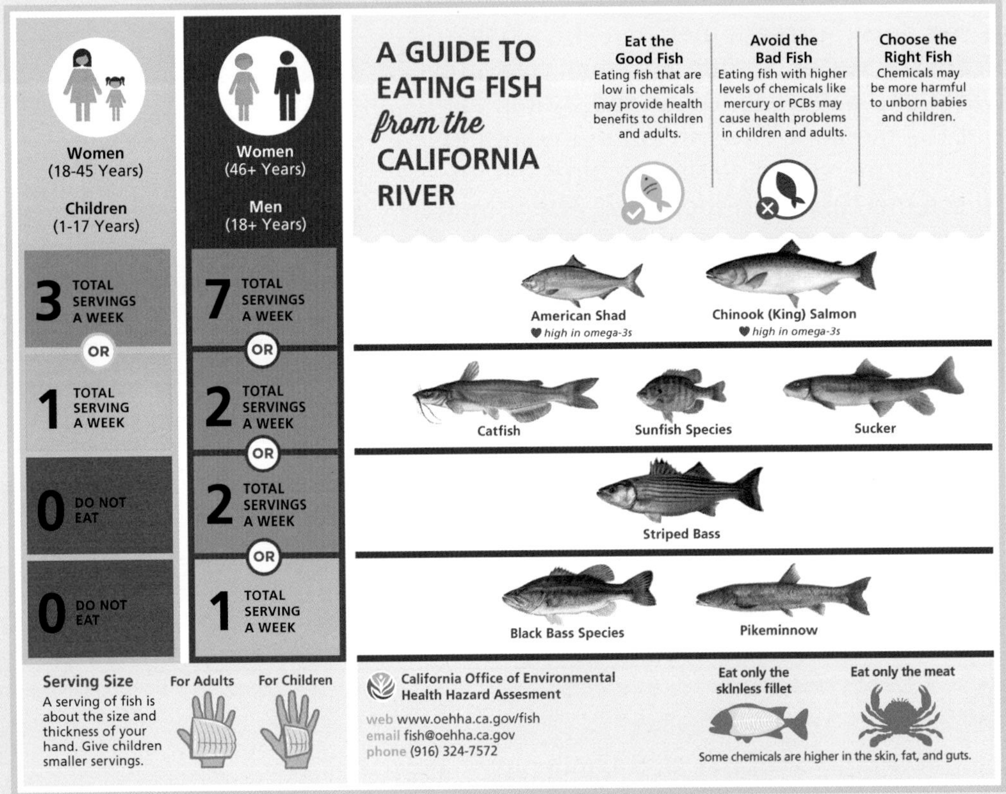


**3.17** *An advisory leaflet for eating different types of fish. There are slightly different recommendations for men and women, and people of different ages. (Reproduced with permission from the California Office of Environmental Health Hazard Assessment.)*

## Key facts:

- ✔ The build-up of a substance in an organism because the substance cannot be broken down and is not excreted is known as bioaccumulation.
- ✔ Some poisonous substances do not break down quickly and so bioaccumulate in organisms.

## Check your skills progress:

- I can model energy flow and bioaccumulation.

# Invasive species

You will learn:

- To describe the effects of new or invasive species on an ecosystem
- To carry out practical work safely
- To collect and record observations and measurements appropriately
- To present and interpret scientific enquiries correctly
- To evaluate secondary information for relevance and potential bias

## Starting point

| You should know that... | You should be able to... |
|---|---|
| The number of a certain type of organism in an area is its population | Draw and interpret food chains and food webs |

## Competition

To survive and reproduce, animals need four key resources:

- water
- food
- shelter
- resources to help reproduction – for example, finding a partner for sexual reproduction.

For plants to survive and reproduce, they usually need:

- water
- gases from the air
- light
- shelter
- a source of mineral salts
- warmth
- resources to help reproduction – for example, insects or wind for pollination.

Organisms living in the same ecosystem often need the same resources. There is **competition** between organisms to get those resources. For example, plants often compete for light, water and mineral salts. Plants that cannot get enough of these resources will die. If a resource is scarce, there will be more competition for it.

### Key term

**competition**: a struggle between some organisms for the same resources.

1 a) Write down *four* resources that an animal needs to survive and reproduce.

b) Explain why a plant needs water to survive.

**Key term**

**food web**: diagram to show how food chains interconnect in a habitat.

We can model the competition for food between animals in an ecosystem using a **food web**.

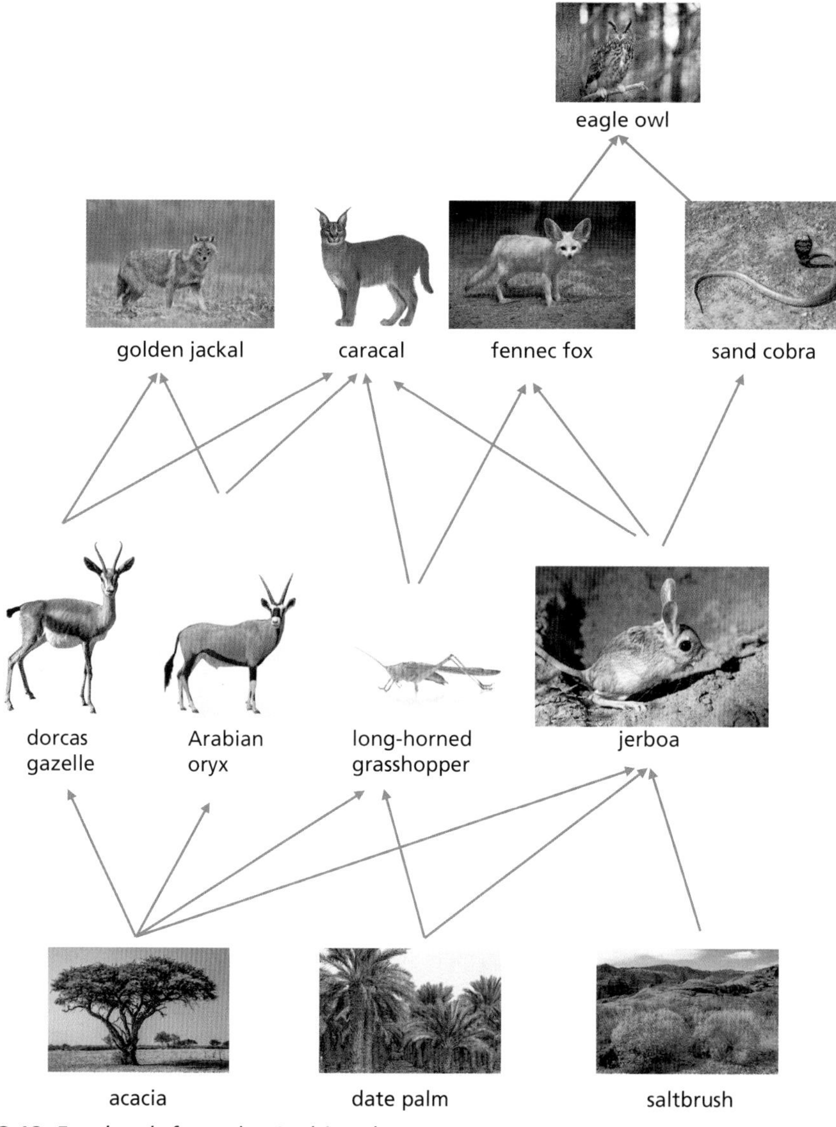


**3.18** *Food web from the Arabian desert ecosystem.*

2 Which organisms compete for:

a) long-horned grasshoppers

b) dorcas gazelles?

3 For what do sand cobras compete with caracals?

Scientists use food webs to predict what happens if there is a change in a population. For example, if the number of grasshoppers goes down, there is less food for fennec foxes. Some may starve and the population of the foxes decreases.

**4** Predict and explain what will happen to the dorcas gazelle population if:

**a)** acacia trees get a disease, which kills many of them

**b)** there is a decrease in the golden jackal population.

**5** If no rain falls on the desert for a few years, explain why competition between the primary consumers will increase.

## Non-native species

The species that are naturally found in an area are **native**. All the organisms in figure 3.17 are native to the Arabian desert.

When a new (**non-native**) organism enters an ecosystem, it can cause problems. For example, golden eagles are not native to the Arabian desert but eat the same animals that eagle owls do. If many golden eagles started to live in the desert, more fennec foxes and sand cobras would be eaten, and there may be fewer eagle owls.

New organisms arriving in an area can cause other problems. They may bring a disease that spreads to other organisms. They may also use up other resources, apart from food. For example, non-native birds may occupy the nesting sites that native birds should be using.

**6** Give the reason why the Arabian oryx is said to be native to the Arabian desert ecosystem.

## Invasive species

An **invasive species** is a non-native species that causes a lot of damage to an ecosystem. Invasive species are often not eaten by things in their new ecosystem, and so their populations grow quickly.

In the 1940s, brown tree snakes were accidentally introduced to the island of Guam in the Pacific Ocean (probably on a boat). The snakes eat birds, and scientists think that they have caused the **extinction** of at least ten native bird species.

### Key terms

**extinction**: when a species dies out completely.

**invasive species**: a non-native species that damages an ecosystem.

**native species**: a species that naturally lives in an area.

**non-native species**: a species that does not naturally live in a certain area.

In the 19th century, farmers planted prickly pear cactus plants in Australia to make hedges. Within a few decades, 40 000 $km^2$ of land was covered with the plants. They used up the water that native plants needed and meant that the land could not be farmed.

Grey squirrels were brought to the United Kingdom (UK) from the USA in the 1870s because people thought they were 'cute'. They eat unripe nuts from trees, but the UK's native red squirrels only eat ripe nuts. In years when the trees do not produce many nuts, the red squirrels starve. Today, red squirrels are uncommon in the UK.

**3.19** *A prickly pear cactus*

**3.20** *Grey squirrels win the competition with red squirrels.*

## Activity 3.3: Researching the effects of invasive species

In this activity you will use secondary sources to research problems caused by invasive species.

**A1** Use secondary sources to identify three invasive species that have caused problems in different countries. (You could use one or more of the examples shown above.)

**A2** Design a table to present the information you collect. Include the name of the species, where it is native, where it has caused problems, the problems it has caused, and what scientists are trying to do about it.

**A3** Record all the sources of information you look at. Choose three of these sources and evaluate them by listing their good and bad points. Write down which was the most useful of these sources and why, and which was the least useful and why.

**A4** You may be asked to present what you have found to your class.

Figures 3.21 and 3.22 show examples of plants that have been introduced into ecosystems and have become invasive species.

**7** Look at figure 3.21.

**a)** Give a reason why kudzu kills trees in forests.

**b)** Give the reason why kudzu is able to spread in this way.

**8** Look at figure 3.22. Algae are unicellular organisms that live in water and photosynthesise. Explain why there are now fewer algae in Lake Naivasha.

**3.21** *Kudzu is a fast-growing vine native to Japan and Southeast China. It is now destroying forests in the USA.*

**3.22** *Water hyacinth, which is native to Brazil, got into Lake Naivasha in Kenya in 1986. Its spread has killed other plants and made the lake difficult to navigate.*

## Key facts:

- ✔ Organisms compete for the resources they need.
- ✔ An invasive species is one that is new to an area and causes damage to the ecosystem.
- ✔ If one organism is growing very well, it may use up the resources that other organisms need to survive.

## Check your skills progress:

- I can use a model to make predictions.
- I can carry out research, evaluate sources and record the information I have found out.

### Making links

In Chapter 2, Topic 2 you may have learned about microorganisms. In a savanna ecosystem, the lions are a top predator. This means that they are the last animal in any food chain. However, lions still provide food for other organisms. Describe how this happens.

# End of chapter review

## Quick questions

1. An ecosystem is:

   **a** all the animals that live in a certain area

   **b** all the organisms that live in a certain area and the physical factors that affect that area

   **c** a set of physical factors that affect an area and all the plants that live in that area

   **d** all the organisms in a continent and all the factors that affect their populations.

2. An example of a habitat is:

   **a** the Atlantic Ocean

   **b** the rainy season

   **c** the Arctic region of the Earth

   **d** underneath dead leaves in a woodland

3. The build-up of a substance inside organisms in a food chain is called:

   **a** biology　　**b** bioaccumulation

   **c** biometrics　　**d** non-biodegradable

4. A native species:

   **a** is naturally found in a certain area

   **b** can only be grown on farms

   **c** has been introduced by people into an area

   **d** has moved into an area from another area

5. Which of these would you use to present information about the populations of different species in an area?

   **a** line graph　　**b** bar chart

6. Look at the following words. Write down all the words that are 'physical factors.'

   cold　desert　forest　fox　grass　lake　wet　windy

7. Scientists take samples from a habitat to find out what is living there. What method of sampling is best for the following organisms?

   a small plants

   b animals that run along the ground

   c small animals that live at the tops of tall grass plants

8. **(a)** State *four* resources needed by animals from their ecosystem.

   **(b)** A non-native species of bird moves into an area. Explain *two* ways in which it might cause the population of native birds to fall.

   **(c)** The population of a non-native species rises very quickly. Give a reason why this can happen.

   **(d)** What name is given to a non-native species whose population increases very quickly?

9. Gazelles eat grass and other small plants. Not all of the energy in their food stays as energy in their bodies. State *two* ways in which energy is lost from an organism.

## Connect your understanding

10. Scientists measured the amount of mercury in some different fish in a bay. Their results are shown in the table.

| Organism | Mercury level (ppm) |
|---|---|
| blackfin tuna | 0.9 |
| grey triggerfish | 0.19 |
| king mackerel | 3.76 |
| red snapper | 0.41 |
| scamp grouper | 0.24 |

   **(a)** Explain why the scientists advised people not to eat some of these fish.

   **(b)** Explain why mercury builds up in the tissues of fish.

   **(c)** Explain which of these organisms is likely to be at the end of a food chain.

   **(d)** What sort of chart or graph would you use to present this data?

11. Look at this food chain.

grass → rabbit → fox

In the food chain, identify:

(a) a producer

(b) a secondary consumer

(c) a herbivore

(d) an organism in the second trophic level

12. A series of pitfall traps was used to study the populations of different invertebrates that lived in dead leaves on a forest floor. The table shows the results. Woodlice eat leaves, ground beetles eat woodlice and wolf spiders eat ground beetles.

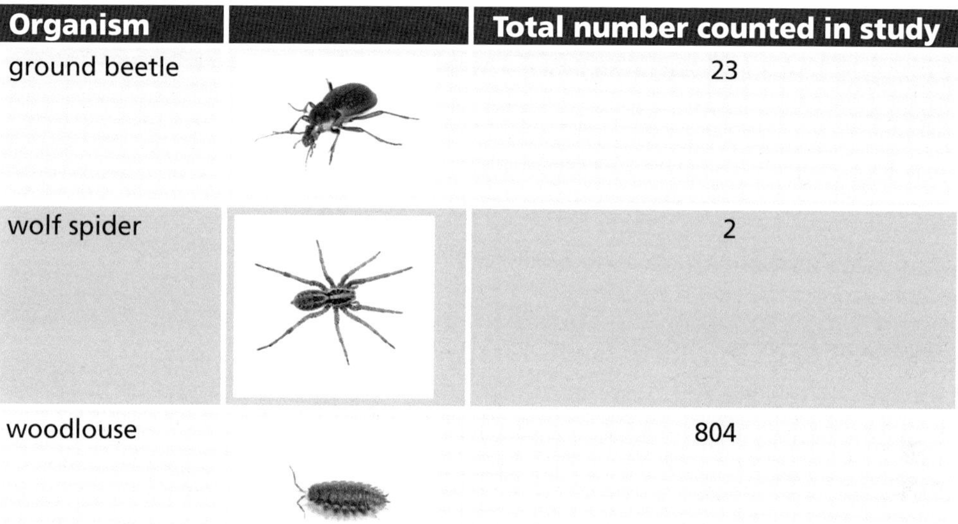

| Organism | | Total number counted in study |
|---|---|---|
| ground beetle | | 23 |
| wolf spider | | 2 |
| woodlouse | | 804 |

(a) Draw a food chain for this habitat.

(b) Explain why the number of organisms changes along the food chain.

13. Scientists took samples in woodland in Turkmenistan. They wanted to find out about the small plants living among the trees. In the results table, the ticks show the plants found in each sample.

| Type of plant | Sample number | | | | | | | | | | | |
|---|---|---|---|---|---|---|---|---|---|---|---|---|
| | 1 | 2 | 3 | 4 | 5 | 6 | 7 | 8 | 9 | 10 | 11 | 12 |
| A – nettle tree (*Celtis*) | | ✓ | | ✓ | | ✓ | | ✓ | | ✓ | ✓ | |
| B – cherry (*Cerasus*) | ✓ | | | | | | ✓ | | ✓ | | ✓ | |
| C – ephedra (*Ephedra*) | | | ✓ | ✓ | | ✓ | | | | ✓ | | |

**(a)** Draw a bar chart to show the number of samples each plant was found in.

**(b)** Which was the most common plant?

**(c)** Why is this information not shown on a line graph?

## Challenge question

14. Kudzu is a fast-growing vine plant that is an invasive species in the USA. It can quickly grow in such a way that it completely covers all the trees.

**(a)** Explain how kudzu can destroy a woodland ecosystem.

**(b)** Kudzu is native to Japan and Southeast China.
Suggest *two* ways in which it might have been spread to the USA by human activities.

## End of stage review

1. The tables show nutritional information from two different foods.

| Sliced bread | Amount per serving (1 slice) |
|---|---|
| Energy | 340 kJ |
| Fats | 4 g |
| Carbohydrates | 12 g |
| Protein | 2 g |

| Cookies | Amount per serving (1 cookie) |
|---|---|
| Energy | 330 kJ |
| Fats | 5.7 g |
| Carbohydrates | 7.2 g |
| Protein | 1 g |

   **(a)** Cedric says that bread contains more energy than cookies. Explain why he cannot make this conclusion.

   **(b)** What information would you need to be able to make a fair comparison between the nutrients in the foods?

   **(c)** State what your body uses carbohydrates for.

2. The diagram shows a knee joint.

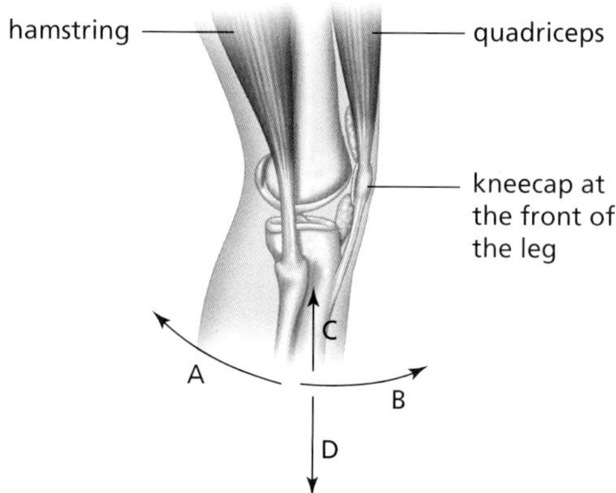


   **(a)** When the quadriceps contracts, in which direction does it move the lower leg bones? Choose from A, B, C or D.

   **(b)** The hamstring moves the lower leg bones in the opposite direction to the quadriceps. What is a pair of muscles like this called?

   **(c)** Why do bones need to be moved by pairs of muscles, rather than single muscles?

**(d)** What sort of joint is the knee joint?

**(e)** Muscle cells need a lot of energy. Explain what small part of cells muscle cells will contain a lot of.

**(f)** Give the name of a nutrient, needed in small amounts, that is needed to build strong bones.

**3. (a)** The diagram shows the human respiratory system.

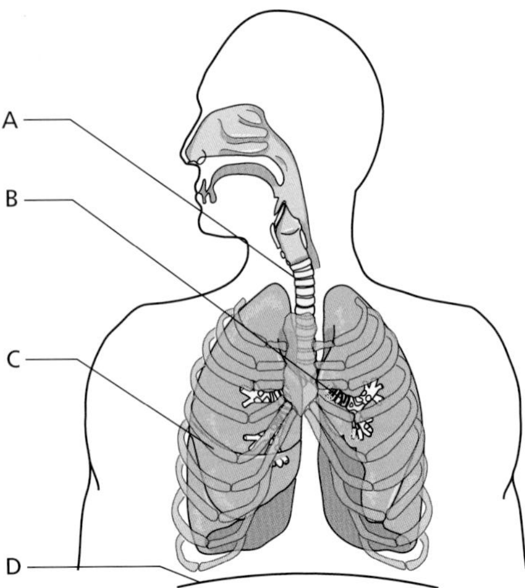


**(i)** Name the parts labelled A and B.

**(ii)** Describe how muscles in part D help when a person breathes in.

**(iii)** Gas exchange happens in the lungs. Copy and complete the following sentence to describe what happens using the best word from the list.

| breathe carry diffuse dilute dive ventilate |
|---|

During gas exchange, oxygen particles ________________ into the blood.

**(iv)** Oxygen is used in aerobic respiration. Complete the word equation for this process.

_____________ + oxygen → carbon dioxide + _____________

**(v)** Smokers are more likely to get lung infections caused by microorganisms. Give the name of the blood cells that will help to destroy these microorganisms.

**(b)** An athlete runs at different speeds on a treadmill (a running machine used in training).
A scientist measures the athlete's heart rate at each speed. The results are shown on the graph.

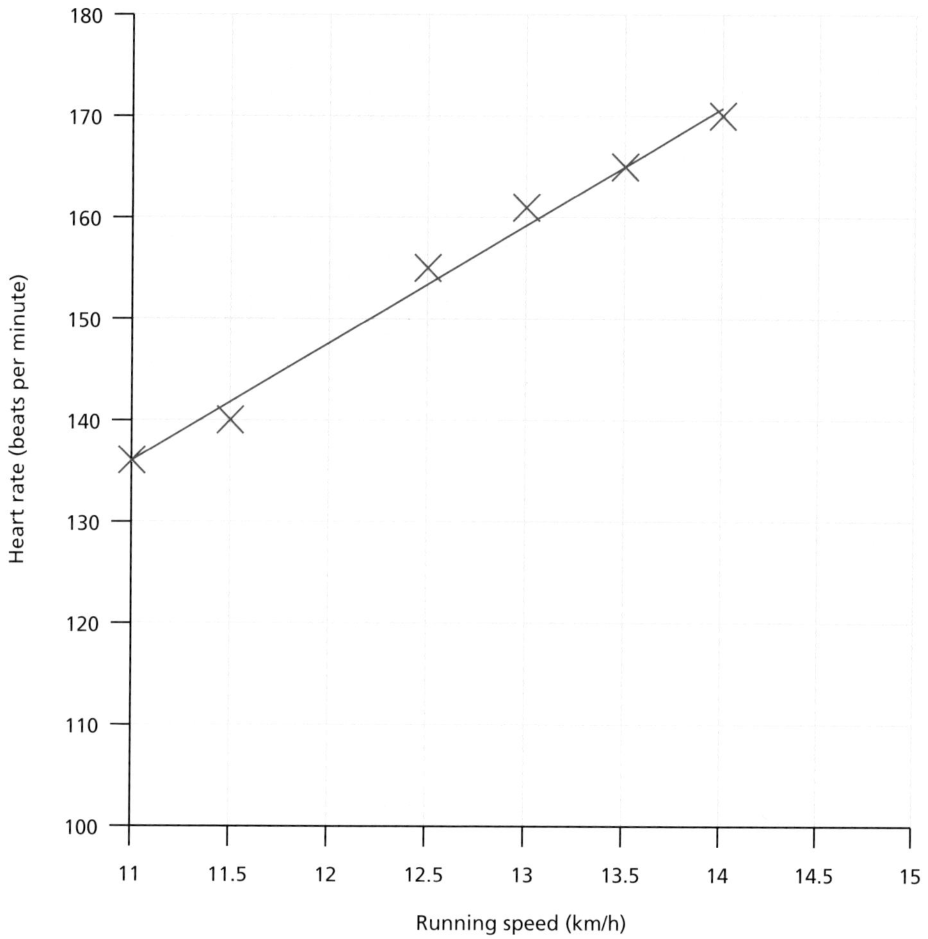


**(i)** Describe the pattern of results shown in the graph.

**(ii)** Explain the link shown in the graph.

**(iii)** Estimate the athlete's heart rate at a speed of 12 km/h.

**4.** The figure shows part of a food web from the Amazon rainforest ecosystem.

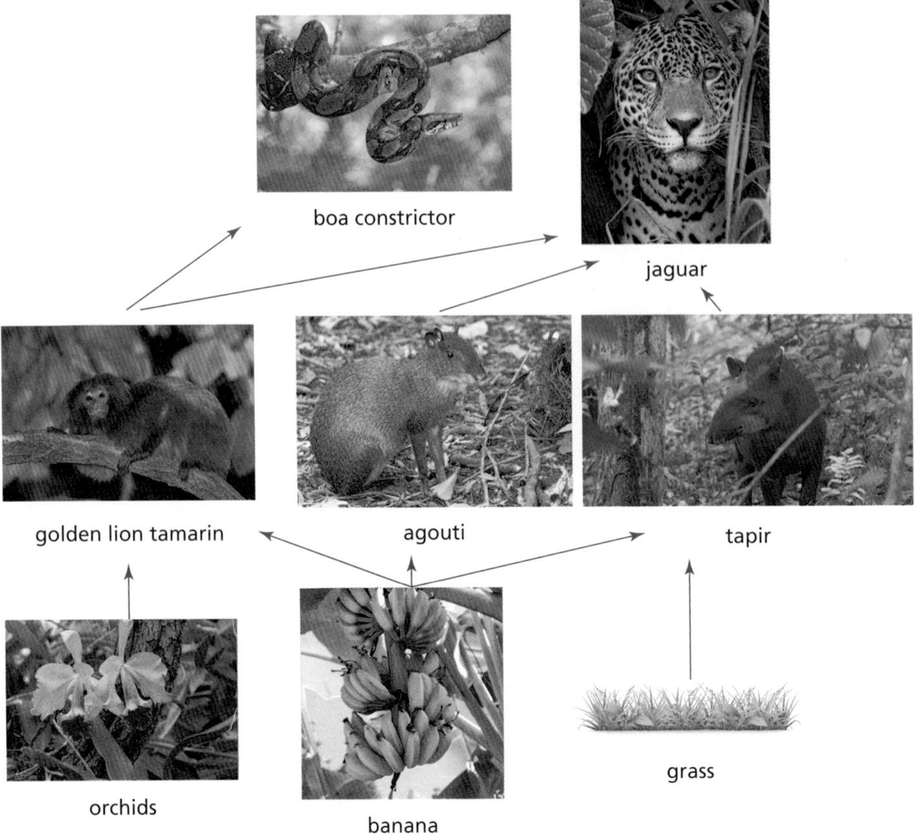


**(a)** **i)** How might a decrease in the population of tapirs affect the population size of jaguars? Explain your answer.

**ii)** Describe *one* difference in the physical factors of the rainforest floor habitat of the agoutis and the tree habitat of the golden lion tamarins.

**(b)** A new species of rat arrives in this ecosystem on a tourist boat. It eats orchids and its population grows so quickly that the ecosystem is harmed.

**i)** Explain the effect that competition from the rat has on the population of *one* other animal in the food web.

**ii)** What name is given to a species, such as the rat, that is not native to the ecosystem?

**(c)** Illegal gold mining in the Amazon rainforest releases poisonous substances into the ecosystem, such as mercury.

**i)** Explain why mercury bioaccumulates in the bodies of agoutis.

**ii)** State the reason why you would expect to find more mercury in jaguars than in agoutis.

**(d)** Give the name of another ecosystem.

# Chemistry


*Chapter 4: Structure and properties of materials*

4.1: Structure of an atom 79
4.2: Paper chromatography 83
End of chapter review 87

*Chapter 5: Solutions and solubility*

5.1: Concentration of solutions 90
5.2: Solubility 94
End of chapter review 97

*Chapter 6: Chemical changes*

6.1: Using word equations 100
6.2: Pure substances and mixtures 105
6.3: Measuring temperature changes 109
6.4: Exothermic and endothermic processes 113
6.5: The reactivity series 119
End of chapter review 126
End of stage review 129

4

## Chapter 4

Structure and properties of materials

### *What's it all about?*

Atoms are the building blocks of the Universe.

Every object in the Universe is made up of atoms from just fewer than 100 different elements, which are listed in the Periodic Table.

Just like toy building blocks, atoms can be arranged to form many different structures.

You will learn about:

- The structure of an atom
- The charges associated with each part of an atom
- How paper chromatography can be used to separate and identify substances in a sample

You will build your skills in:

- Describing analogies and how they can be used as a model
- Using symbols to represent scientific ideas
- Making conclusions by interpreting results
- Explaining the limitations of conclusions

Chapter 4 . Topic 1

# Structure of an atom

You will learn:

- To describe the Rutherford model of the atom
- To know the charges on electrons, protons and neutrons
- To know how individual atoms are held together
- To describe how evidence affects scientific hypotheses
- To explain an analogy and how to use it as a model
- To use analogies

## Starting point

| You should know that... | You should be able to... |
|---|---|
| All matter is made from atoms | Make predictions based on scientific understanding |
| Elements only contain one type of atom | Make conclusions by interpreting results |
| Ideas can be tested by carrying out investigations and collecting evidence | Understand the strengths and limitations of a model |

## Changing models

Everything is made up of atoms so understanding what is inside an atom helps us to work out how atoms join together to form different substances. But atoms are far too small to be seen. Scientists have carried out experiments on groups of atoms and have used the evidence to produce a **model** of what is inside an atom. This helps us to understand what atoms are like.

Models may change over time because scientists find new evidence that the old model cannot explain. When a model explains a difficult idea by comparing it to something more familiar, it is known as an **analogy**.

### Key terms

**analogy**: a type of model that compares something unfamiliar to something more familiar

**model**: simple way of showing or explaining a complicated object or idea based on evidence.

## First ideas about atoms

Evidence for the existence of atoms became available during the 18th and 19th centuries when scientists studied the properties of elements and how they react together. They thought that atoms were hard spheres, so small that you couldn't break them up into even smaller parts (see figure 4.2).

## Activity 4.1: The mystery box

Work in teams of three or four. Your teacher will give you a mystery box. Your task is to work out what is inside *without opening it*.

You will need to think about what experiments you can do that will provide evidence about what is inside.

**A1** What experiments did you use?

**A2** What evidence did you collect?

**A3** What did the evidence tell you about what is inside?

**A4** Did you change your ideas after each experiment? If so, explain why.

**A5** How sure are you that your idea is correct? Give a reason for your answer.

**4.1** *How can you find out what is inside?*

This idea changed in 1904 when J.J. Thomson did experiments to show that atoms contain particles that have a negative charge. He called them **electrons**. Because atoms are neutral he concluded that there must also be positively charged parts in the atom. His model of the atom was called the **plum pudding model** because the electrons seemed to be spread out in a cloud of positive charge. The electrons were like the plums and the rest of the atom was the pudding. This is an example of an analogy.

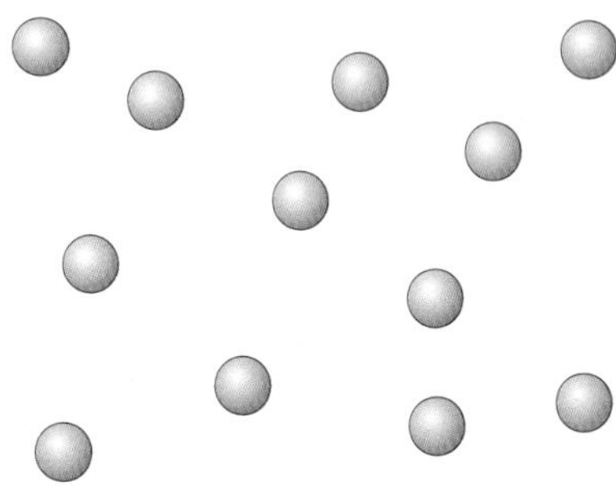

**4.2** *An early model of atoms.*

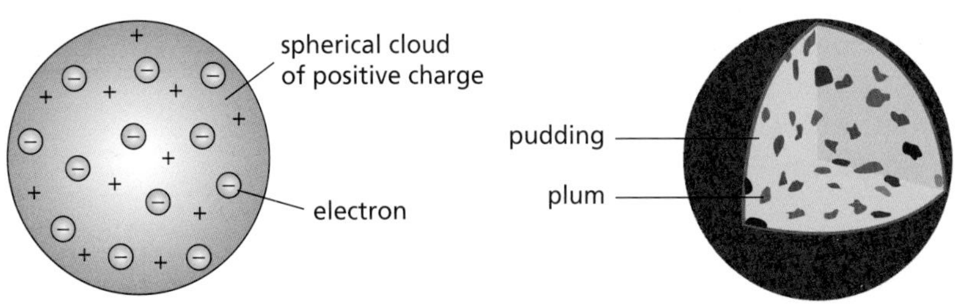


**4.3** *Thomson's plum pudding model was named after an old English pudding containing plums (a type of fruit)*

## Rutherford's nuclear model

In 1909 a team of scientists, led by New Zealand scientist Ernest Rutherford, fired small, fast-moving **alpha particles** at thin sheets of gold foil. Rutherford had already proved that alpha particles, emitted by some radioactive materials, had a positive charge and were smaller than an atom. He also knew that two positive charges repel each other.

If the plum pudding model was correct, the alpha particles should pass straight through the gold foil.

### Key terms

**alpha particle**: particle with a positive charge given out by some radioactive elements – it is smaller than an atom.

**electrons**: very small negatively charged particles in an atom.

**plum pudding model**: early model of the atom – a cloud of positive charge with electrons embedded in it.

Most of the particles did this, but a few were deflected – some were even bounced back. This was an unexpected result – it did not match Rutherford's predictions.

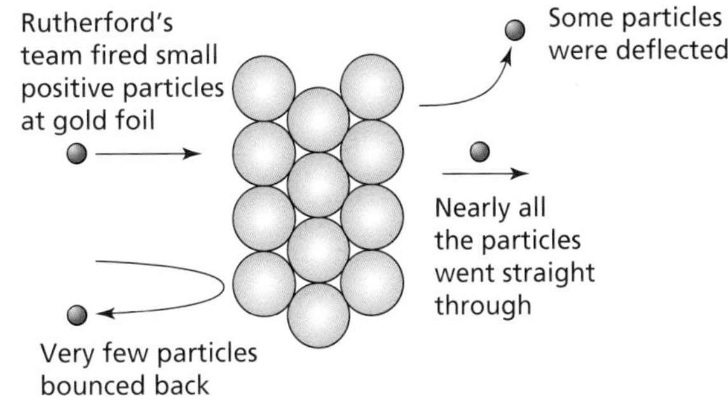


**4.4** *The gold foil experiment.*

Rutherford changed the model to fit this new evidence. He suggested that the atom had a small, positively charged **nucleus** with the electrons surrounding the nucleus in a 'cloud'. He thought that the atom was mostly empty space. This was a big change from previous models, which depended on the idea that atoms were solid.

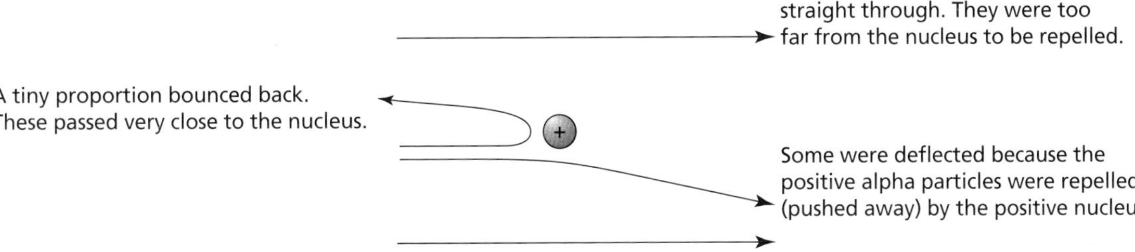


**4.5** *Explaining the gold foil experiment.*

1. Explain why the plum pudding model of the atom had to change after Rutherford's gold foil experiment.
2. Explain how Rutherford's experiment proved that the nucleus is positively charged and very small compared to the rest of the atom.

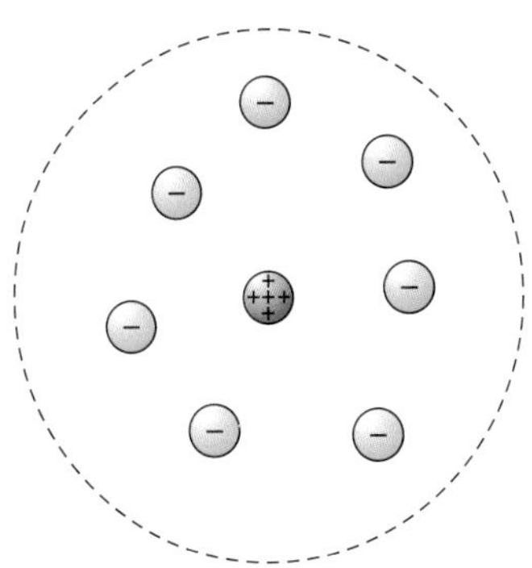

**4.6** *Rutherford's model of the atom.*

## Improving the model further

Scientists continued work to improve Rutherford's model. Experiments in the 1920s and 1930s showed that the nucleus was made up of many individual parts – positively charged **protons** and those with no charge called **neutrons**.

| Particle | Charge |
|---|---|
| Electron | Negative |
| Proton | Positive |
| Neutron | Neutral – no charge |

**Table 4.1** *Charges of the different particles.*

The nucleus is very small, it is about 10 000 times smaller than the size of the whole atom. It is difficult to imagine this, so you can use a model to help you.

### Key terms

**neutrons**: particles with no charge in the nucleus of an atom.

**nucleus**: the central part of an atom – contains protons and neutrons.

**protons**: positively charged particles in the nucleus of an atom.

Imagine that the nucleus of an atom is the size of a pea. The whole atom would be the same size as a large sports stadium.

3 Suggest why the neutron was the last particle to be discovered.

## Electrostatic attraction

Opposite charges attract. It is the attraction between the positively charged nucleus and the negatively charged electrons that holds an atom together. This is called electrostatic attraction.

4 What particles make up the nucleus?

5 Why does the nucleus have a positive charge?

### Key facts:

- ✔ Rutherford's model was that an atom has a small, central, positively charged nucleus, but most of it is empty space.
- ✔ Electrons have negative charge, protons have positive charge and neutrons have no charge.
- ✔ The electrostatic attraction between positive and negative charges holds together individual atoms.

### Check your skills progress:

- I can make predictions of likely outcomes based on scientific knowledge.
- I can describe an analogy and how it can be used as a model.

# Paper chromatography

You will learn:

- To describe the use of paper chromatography with substances
- To describe results in terms of any trends and patterns and identify any abnormal results
- To reach conclusions studying results and explain their limitations
- To describe the application of science in society, industry and research

## Starting point

| You should know that... | You should be able to... |
|---|---|
| A mixture is made up of at least two different elements or compounds | Make conclusions by interpreting results |
| Substances in a mixture can easily be separated | |
| When a solid dissolves in a liquid the solid is still present | |

## Paper chromatography

The substances that make up a mixture have different chemical and physical properties.

These differences mean that it is easy to separate them.

**Chromatography** is used to separate mixtures of **soluble** substances, often coloured substances like the dyes in ink or food colouring. It works because the soluble substances have different properties.

You can use paper chromatography to separate the dyes in ink.

First, you use a pencil to draw a line across the bottom of a strip of filter paper. Next, you put small spots of the ink on the line.

The paper is put into a beaker of a **solvent**, like water, so the pencil line is just above the surface of the solvent. The solvent travels up the paper. After a while you will see coloured marks as the individual dyes in the ink are separated out. The final pattern is called a **chromatogram.**

### Key terms

**chromatogram**: the pattern of spots produced during chromatography.

**chromatography**: a technique used to separate soluble substances (usually coloured dyes or inks).

**soluble**: substance dissolves to form a solution.

**solvent**: a liquid that dissolves a soluble substance.

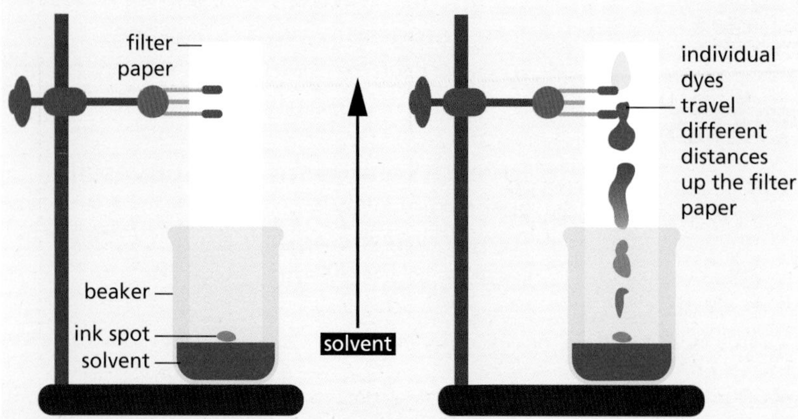


**4.7** *Equipment used for paper chromatography.*

## How chromatography works

If the dyes in the ink are soluble, they will **dissolve** in the solvent.

They will then travel up the paper because they are attracted to both the solvent and the paper.

The dyes will separate out because there are differences in how much each is attracted to the paper and the solvent. This affects how far they travel up the paper.

### Key term

**dissolve**: when a soluble substance becomes a solution.

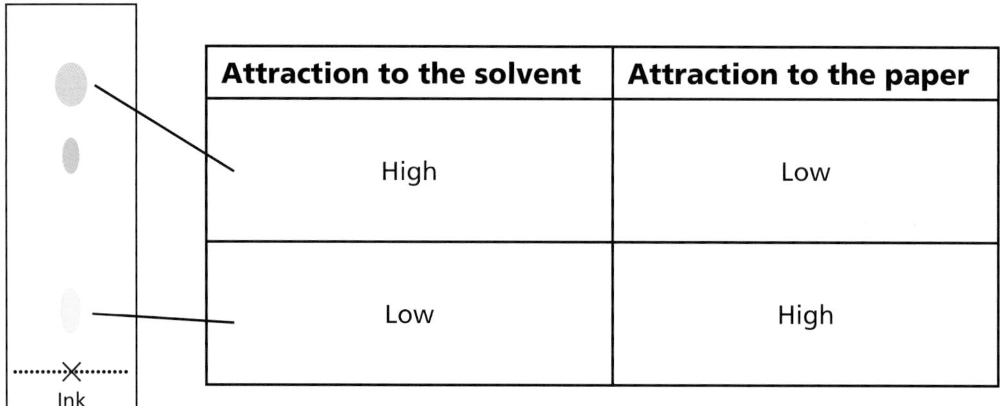


| Attraction to the solvent | Attraction to the paper |
| --- | --- |
| High | Low |
| Low | High |

**4.8** *The position of the dye depends on its attraction to the solvent and to the paper.*

1. How many dyes are in the ink shown in the chromatogram in figure 4.8? Explain how you can tell.
2. Desi used chromatography to separate the ink in a pen. The ink spot did not move up the paper. Suggest why.

**4.9** *Sweets containing food dye.*

## Using chromatography

Chromatography can be used to identify unknown substances in a mixture. This has many uses.

A food scientist uses chromatography on the coloured covering of some sweets. She also adds different food colourings to the paper.

She compares the spots produced by each.

The chromatogram shows that the sweets must contain food colourings E102 and E133 because spots are seen in the same places.

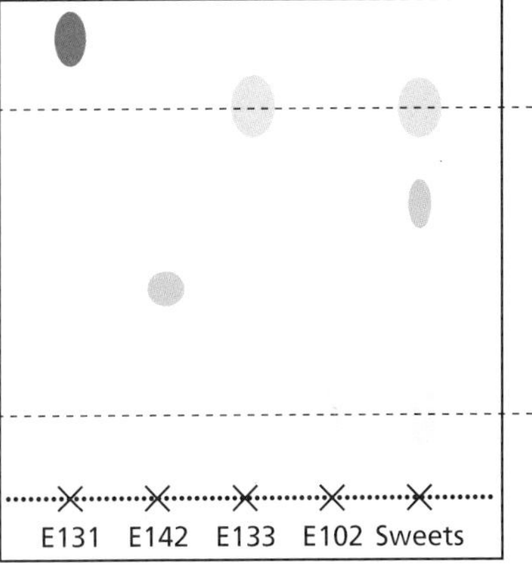


**4.10** *Chromatogram of the food colouring from the sweets.*

**3** Do the sweets contain any E131? Explain how you can tell, using figure 4.10.

**4** Suggest why the spot of E102 from the sweets is larger than the spot from the sample of E102.

**5** Suggest why the scientist drew lines across the centre of each spot.

## Science in context: Applying chromatography

Chromatography is a useful technique used by scientists in many different jobs across the world.

It is particularly important in forensics – a type of science used to help the police solve crimes.

Inks, paints or dyes at crime scenes are collected and compared to known samples in order to identify them. Sometimes car paint is found scraped on a building or another car at a crime scene. Chromatography is used to identify the paint. This can be used to identify the make of the car, which can be used as evidence to help solve the crime.

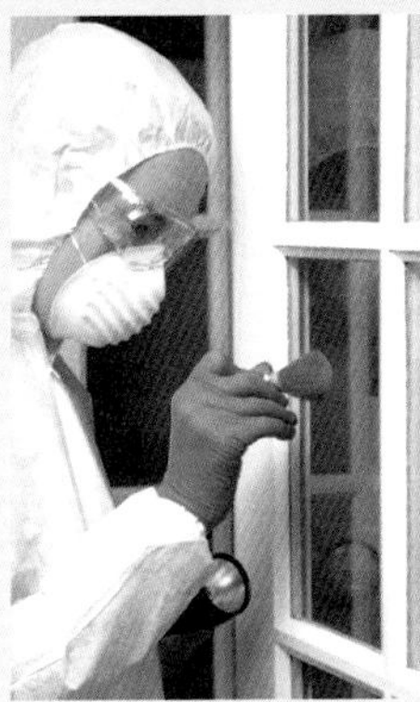

**4.11** *Collecting evidence at a crime scene.*

## Activity 4.2: Applying chromatography

Your teacher will show you a chromatogram produced using chromatography of ink found at a crime scene. You have pens taken from the three suspects.

**A1** Use chromatography to separate the dyes in the ink from the three pens.

**A2** Use your results as evidence to identify which pen was used at the crime scene.

**A3** Usually a real chromatogram doesn't look like the ones in figure 4.10. Sometimes the ink spots overlap, or clear spots are not produced. Describe:

- **a)** if you had any of these issues
- **b)** why they were a problem when it came to using your results as evidence
- **c)** what you did to solve this.

## Key facts:

- ✔ Chromatography is used to separate mixtures of soluble substances.
- ✔ The final pattern created is called a chromatogram.
- ✔ During chromatography soluble dyes separate out at different locations.
- ✔ Chromatography can be used to identify unknown substances in a mixture.

## Check your skills progress:

- Make conclusions by interpreting results.
- Explain the limitations of the conclusions.

### Making links

Topic 6.2 introduces purity. It is very important that the medicines produced by chemical reactions are pure (not a mixture). If a scientist used a chemical reaction to make a medicine, how could they use chromatography to determine if it is pure?

# End of chapter review

## Quick questions

1. Describe the Rutherford model of the structure of an atom.
2. Give the charges (positive, negative or neutral) for electrons, protons and neutrons.
3. What *two* particles are found in the nucleus of an atom?
4. Describe why the nucleus has a positive charge.
5. Name the force that holds the electrons to the nucleus in an atom.
6. Which mixture could be separated by chromatography?
   - **(a)** Salt and sand
   - **(b)** Different coloured dyes
   - **(c)** Water and ethanol

## Connect your understanding

7. A carbon atom has no overall charge. It contains six protons. How many electrons does it have? Explain how you worked this out.
8. Explain why the electrons are attracted to the nucleus of an atom.
9. The gold foil experiment performed by Rutherford and his team was repeated many times by them and by other scientists. Suggest why this is important.
10. The nucleus is about 10 000 times smaller than the size of the whole atom. Salasi used a large glass bead to represent the nucleus. It has a diameter of 1 cm. Calculate how big the whole atom would need to be in metres (100 cm = 1 m).
11. The diagram shows a chromatogram. Which substance (A or B) is pure? Give a reason for your answer.

12. Luca tried to use chromatography to separate the dyes in black ink. He put a dot of the ink on the bottom of some chromatography paper and placed the paper in water. The ink spot did not separate out.

    **(a)** Suggest a reason why.

    **(b)** Describe what he should change so chromatography works.

## Challenge question

13. The outer shell of coloured sweets is dyed with food colouring. Describe the method a scientist should use to find out if a coloured sweet contains a banned food colouring.

5

# Chapter 5

## Solutions and solubility

### What's it all about?

Solutions are all around us – coffee, petrol and rainwater are all examples of solutions. Solutions are also in us – your blood contains a watery solution called plasma. Plasma contains dissolved food particles, carbon dioxide and blood cells.

You will learn about:

- What is meant by the concentration of a solution
- How the solubility of different salts varies with temperature

You will build your skills in:

- Planning investigations and considering the variables
- Describing trends and patterns in results
- Identifying anomalous results
- Presenting measurements appropriately

# Concentration of solutions

You will learn:

- To understand what is meant by the concentration of a solution
- To plan investigations, including fair tests, while considering variables

## Starting point

| You should know that... | You should be able to... |
|---|---|
| All matter is made from atoms | Use the particle model to represent elements, compounds and mixtures |
| Dissolving occurs when a soluble substance is added to a solvent to produce a solution | |

## Solutions

A **solution** is a mixture of a solute and a solvent.

The **solute** is **soluble,** it dissolves into the solvent, forming the solution.

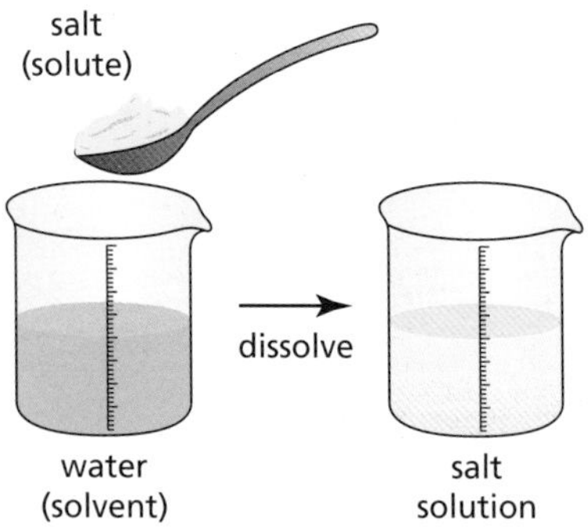


**5.1** *How a salt solution is made.*

Some substances do not dissolve in a solvent. They are **insoluble.**

For example, chalk does not dissolve in water.

### Key terms

**insoluble**: a substance that does not dissolve.

**solute**: a substance that dissolves in the solvent e.g. salt.

**soluble**: substance that dissolves to form a solution.

**solution**: a mixture of a soluble substance and a liquid.

**solvent**: a liquid that dissolves a soluble substance.

**1** Soda water contains bubbles of carbon dioxide gas dissolved in water.

**a)** Name the solvent.

**b)** Name the solute.

**2** Nadia uses acetone to remove her nail varnish.

**a)** What is the solute?

**b)** What is the solvent?

**c)** Why doesn't nail varnish come off in water?

**5.2** *Chalk is insoluble in water.*

## Using a model to explain dissolving

During dissolving the particles in the solute break away from one another. They mix with the solvent to fill the gaps in between the solvent particles.

Figure 5.3 is a model that shows this.

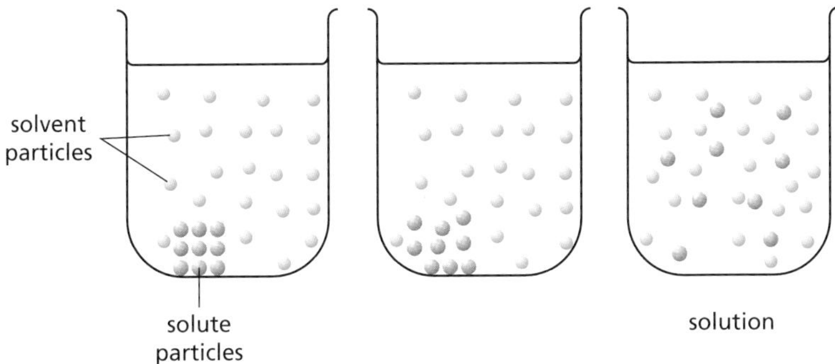


**5.3** *Model of what happens when a solute dissolves.*

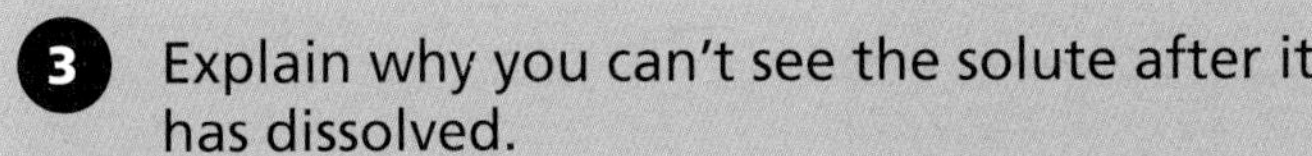

**3** Explain why you can't see the solute after it has dissolved.

## Concentration of a solution

You can make a fruit drink by dissolving powder into water.

The more powder you add, the more concentrated the drink becomes. You can change the **concentration** of the solution.

**5.4** *Adding more powder causes the drink to become more concentrated.*

### Key term

**concentration**: a measurement of how many particles of a certain type there are in a volume of liquid or gas.

A **dilute solution** has a small number of solute particles dissolved in a large **volume** of solvent.

A **concentrated solution** has a large number of solute particles dissolved in a small volume of solvent.

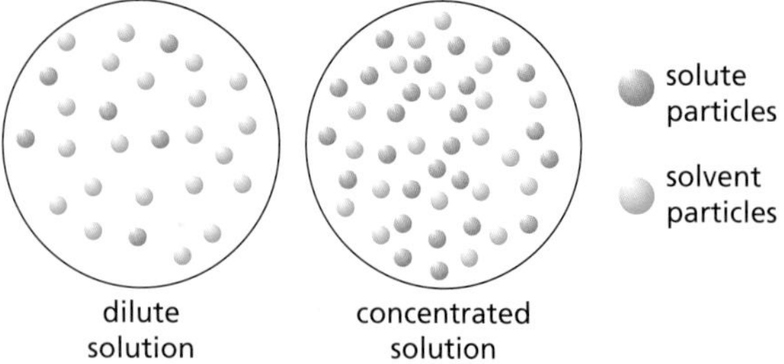


**5.5** *Particle diagrams show the difference between a dilute and concentrated solution.*

### Key terms

**concentrated solution**: a solution that has a large number of solute particles dissolved in a small volume of solvent.

**dilute solution**: a solution that has a small number of solute particles dissolved in a large volume of solvent.

**volume**: how much space an object takes up. Measured in $cm^3$.

**4** Rosa made a cup of black coffee. She decides it is too strong so adds some more water.

a) What happens to the colour of the coffee as she adds more water?

b) Why does this happen?

**5** Davin made three different sugar solutions:

X: 5 g of sugar, 100 $cm^3$ of water

Y: 20 g of sugar, 200 $cm^3$ of water

Z: 25 g of sugar, 50 $cm^3$ of water

a) Which solution is the most concentrated?

b) Which solution is the most dilute?

There is a limit to how much solute will dissolve in a solvent. When no solute will dissolve, it has reached its saturation point and is known as a saturated solution.

### Activity 5.1: Investigating the saturation point of a solute

The saturation point is the maximum mass of a solute that will dissolve in 100 $cm^3$ of water.

**A1** Plan a method that you can use to find out the saturation point of salt (sodium chloride) in cold water.

**A2** Use the model in figure 5.5 to explain why solutes have a saturation point.

## Key facts:

- ✔ The concentration of a solution relates to how many particles of the solute are present in a volume of the solvent.
- ✔ A dilute solution has a small number of solute particles dissolved in a large volume of solvent.
- ✔ A concentrated solution has a large number of solute particles dissolved in a small volume of solvent.

## Check your skills progress:

- I can use an existing model for another purpose.
- I can plan an investigation.

Chapter 5 . Topic 2

# Solubility

You will learn:

- To describe the effects of temperature on the solubility of salts
- To describe results in terms of any trends and patterns and identify any abnormal results
- To present and interpret scientific enquiries correctly
- To describe the application of science in society, industry and research

## Starting point

| You should know that... | You should be able to... |
|---|---|
| Dissolving occurs when a solute is added to a solvent to produce a solution | Use the particle model to represent elements, compounds and mixtures |
| Temperature affects solids dissolving in liquids | Identify anomalous results |

## Solubility

**Solubility** is a physical property of a substance.

Each solute has a different solubility. The table shows the solubilities of some solutes (**salts**) in 100 cm$^3$ of water at 20 °C.

| Solute | Solubility in g |
|---|---|
| Iron(II) chloride | 52.6 |
| Lead(II) iodide | 0.07 |
| Sodium chloride | 36 |

**Table 5.1** *Solubility of different solutes.*

### Key terms

**salt**: A type of compound that consists of metal atoms joined to non-metal atoms, e.g. sodium chloride.

**solubility**: The mass of solute that will dissolve in a volume of solvent at a certain temperature.

1 Which salt from the table is the:

a) most soluble?

b) least soluble?

2 How much sodium chloride will dissolve in 50 cm$^3$ of water at 20 °C?

## Effect of temperature on solubility

When you make iced coffee you add coffee to boiling water and then add ice.

Why don't you just add the coffee to cold water?

The reason is that the coffee is not very soluble in cold water. The solubility of most solutes increases with temperature.

**5.6** *Iced coffee is made from a solution of coffee and water.*

We can use the particle model to explain why temperature affects solubility.

Solutes dissolve because the solvent particles bump into the solute, breaking it up into individual particles.

When the temperature of the solvent increases the solvent particles have more energy, so move around more. This means they hit the solute more often, and with more force.

**3** Figure 5.7 shows the solubility curve for carbon dioxide in water.

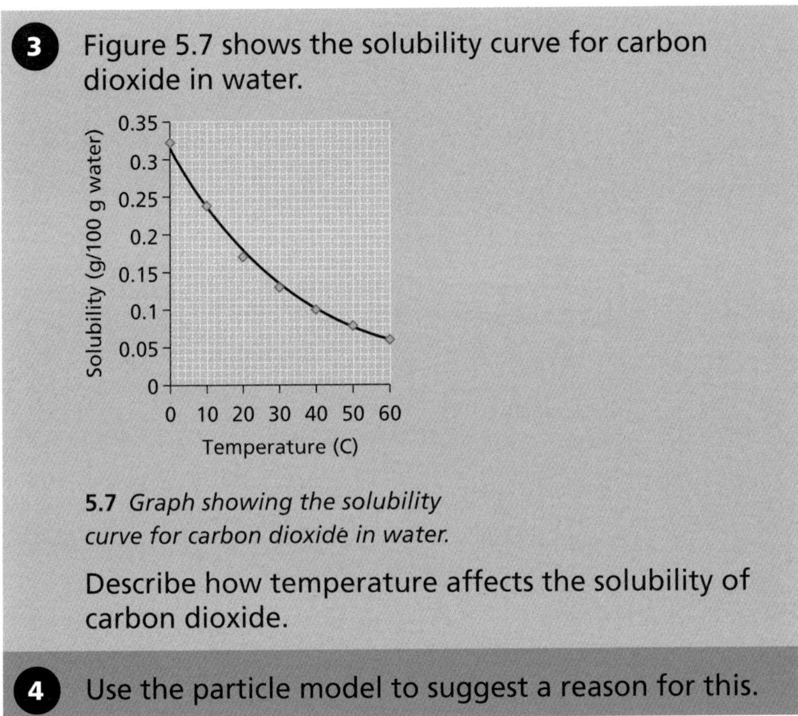


**5.7** *Graph showing the solubility curve for carbon dioxide in water.*

Describe how temperature affects the solubility of carbon dioxide.

**4** Use the particle model to suggest a reason for this.

## Science in context: Soluble drugs

When you take a tablet, the drug dissolves in your digestive system and then passes into your blood. It is carried in your bloodstream to the place in your body where it has an effect and you feel better. There is a problem though – many substances that are useful drugs are insoluble in water. To solve this, scientists have developed a technique that turns an insoluble substance into a soluble substance. This will allow many more drugs to be developed.

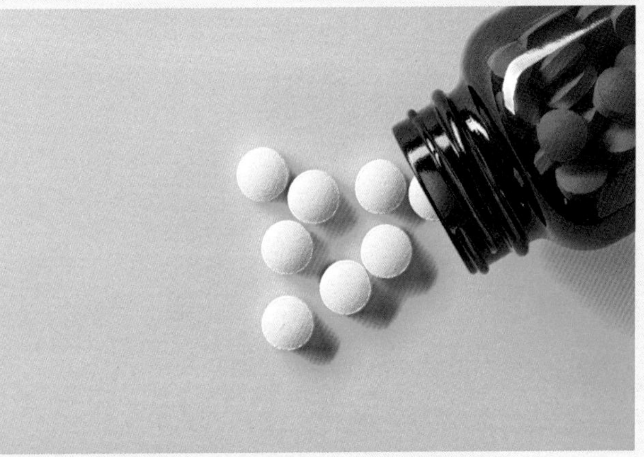

**5.8** *Tablets contain a small amount of a drug.*

## Key facts:

✔ The solubility of salts varies, depending on their type.

✔ The solubility of most salts increases with temperature.

## Check your skills progress:

- I can describe the trends and patterns in results, including identifying any anomalous results.
- I can present measurements appropriately, using graphs.

### Making links

In Topic 1.2 you may have learned that carbon dioxide is carried around the body in the blood. Why is it important that carbon dioxide is soluble in blood?

# End of chapter review

## Quick questions

1. What is a solution?
2. Name the solute and solvent present in saltwater.
3. Describe how the particles are arranged in a concentrated solution.
4. A student made a solution with a weak concentration.
   How can they make it more concentrated?
5. What is meant by the term solubility?
6. What effect does the increase in temperature normally have on the solubility of salts?

## Connect your understanding

7. Draw and label the particles in a concentrated solution and those in a dilute solution.
8. Laila and Ahmed both made copper sulfate solution.

   Laila dissolved 2 g of copper sulfate in 100 $cm^3$ of water.

   Ahmed dissolved 5 g of copper sulfate in 200 $cm^3$ of water.

   Whose solution is the most concentrated? Explain why.
9. Explain what happens when a solute dissolves in a solvent.
10. A maximum of 36 g of sodium chloride will dissolve in 100 $cm^3$ of water at 20 °C.
    How much sodium chloride will dissolve in:

    **a)** 50 $cm^3$ **b)** 200 $cm^3$?

**11.** The graph shows a solubility curve for potassium nitrate.

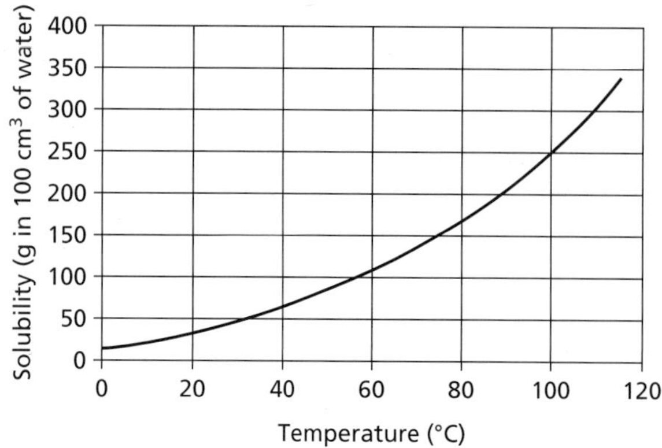


**a)** How much potassium nitrate will dissolve in 100 $cm^3$ of water at 100 °C?

**b)** Amelia wants to dissolve 150 g of potassium nitrate in 100 $cm^3$ of water. She uses water with a temperature of 60 °C. Will she be able to make her solution? Explain why.

## Challenge question

**12.** The graph is a solubility curve for a substance called alum.

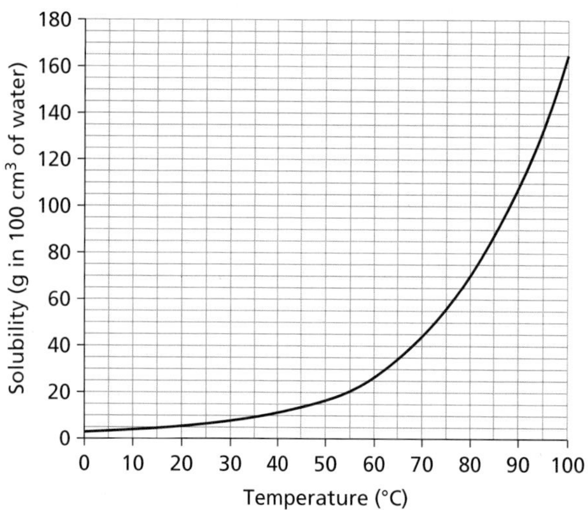


Rohan makes a saturated alum solution using water at a temperature of 80 °C. He then cools the solution down to 40 °C. Describe what will happen and why.

# Chapter 6

## Chemical changes

### What's it all about?

From lift-off to colourful explosions of light, many different chemical reactions take place after a firework is lit.

To lift the firework a powder burns. This reaction releases gases very quickly in a fast reaction, at the same time energy is transferred to the surroundings as heat. The colours come from compounds called salts whose atoms release coloured light when heated. Different metals in the salts give different colours: sodium salts produce yellow sparks and copper salts produce blue.

You will learn about:

- Using word equations to describe chemical reactions
- How purity is a way to describe how much of a chemical is in a mixture
- How reactions do not always lead to single pure substances, but can produce impure mixtures
- Identifying types of reactions by the temperature change measured
- How to determine whether a process or chemical reaction is exothermic or endothermic
- How to describe the reactivity of metals with oxygen, water and dilute acids

You will build your skills in:

- Making predictions of likely outcomes, based on scientific understanding
- Planning a range of investigations, while considering variables
- Choosing equipment and using it appropriately
- Making risk assessments to identify and control risks
- Describing the accuracy of predictions, based on results and suggesting why they were or were not accurate
- Evaluating an investigation and suggesting improvements

Chapter 6 . Topic 1

# Using word equations

You will learn:

- To describe reactions using word equations
- To understand what unreactive substances are
- To choose experimental equipment and use it correctly
- To discuss the global environmental impact of science

## Starting point

| You should know that... | You should be able to... |
|---|---|
| A compound is made up of two or more elements | Use chemical symbols to represent elements |
| A chemical reaction has occurred when there is a loss of reactants or formation of products | Consider hazards when planning practical work |
| The particle model can be used to describe chemical reactions | |

## Chemical reactions

In a chemical reaction the substances that react are called **reactants**. New substances called **products** are made in the reaction.

There are clues that you can look for that show a chemical reaction might have happened. For example:

- Bubbles being given off. This shows that a gas has been made.
- A change in colour. A new product has been made which is a different colour to the reactants.

**6.2** *The statue of liberty was once shiny brown copper but over time this has reacted with substances from the air and has changed colour.*

### Key terms

**product**: substance made during a chemical reaction.

**reactant**: substance that changes in a chemical reaction to form products.

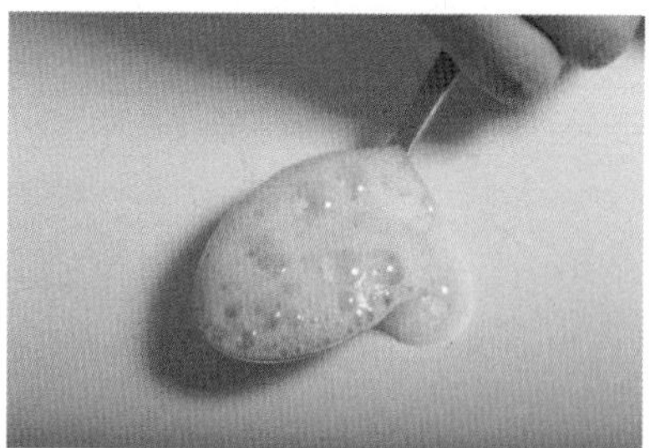

**6.1** *Vinegar reacts with baking powder to produce bubbles of gas.*

**6.3** *The reaction between iron oxide and aluminium releases a lot of heat energy.*

- A change in temperature. This could be an increase or decrease in temperature.

One example of a chemical reaction is rusting.

The reactants are iron, water and oxygen. The product is rust. Iron is a grey metal, rust is brown. This change in colour is evidence that a chemical reaction has taken place.

Some substances are unreactive. They do not take part in chemical reactions. These can be described as **inert**. The noble gases, including helium and neon, are inert because they do not react to form new products.

## Writing word equations

A **word equation** shows us the reactants and products in a reaction. The reactants and products can be elements or compounds.

Magnesium and oxygen react together in an **oxidation** reaction. The product is magnesium oxide. This can be shown in a word equation:

magnesium + oxygen → magnesium oxide

Reactants Product

Word equations always have the reactants on the left. The arrow points to the products that are made.

1. A teacher reacts sodium and chlorine together. Sodium chloride is made. Write a word equation to show this reaction.
2. Photosynthesis is a chemical reaction that takes place in plant leaves. Carbon dioxide and water are used to make oxygen and glucose. Write this as a word equation.

## Naming compounds

There are some rules to follow to help you to name a new compound formed in a chemical reaction.

- If the compound contains a metal, the name of the metal comes first.
- When two elements react to form a compound, the name often ends in *-ide*. For example, iron **chloride** and calcium **oxide**.

### Key terms

**carbonate**: compound that contains carbon, oxygen and another element, for example, calcium carbonate ($CaCO_3$).

**chloride**: compound that is formed when chlorine reacts with another element, for example sodium chloride (NaCl).

**inert**: a chemical that is unreactive.

**oxidation**: chemical reaction with oxygen to form a compound that contains oxygen.

**oxide:** compound that is formed when oxygen reacts with another element; for example magnesium oxide (MgO).

**sulfate**: compound that contains sulfur, oxygen and another element, for example copper sulfate ($CuSO_4$).

**word equation**: model showing what happens in a chemical reaction, with reactants on the left of an arrow and products on the right.

If a compound ends in *-ate* it also contains oxygen. For example, copper **carbonate** contains copper, carbon and oxygen atoms. Calcium **sulfate** contains calcium, sulfur and oxygen atoms.

Metal **hydroxide** compounds only contain atoms of the metal, hydrogen and oxygen.

Table 6.1 shows some examples.

| Name of compound | Metal element | Non-metal element/s | Formula |
|---|---|---|---|
| sodium chloride | sodium | chlorine | NaCl |
| magnesium oxide | magnesium | oxygen | MgO |
| potassium hydroxide | potassium | oxygen, hydrogen | KOH |
| calcium carbonate | calcium | carbon, oxygen | $CaCO_3$ |
| lithium sulfate | lithium | sulfur, oxygen | $Li_2SO_4$ |
| silver nitrate | silver | nitrogen, oxygen | $AgNO_3$ |

**Table 6.1** *The name of a compound usually shows what atoms it contains.*

Remember, if there are more than one atom of an element present in a compound, the number of atoms must be shown in the formulae using sub-script numbers. For example, in the formula $CaCO_3$ there is one Ca atom, one C atom and three O atoms. In the formula $Li_2SO_4$ there are two Li atoms, one S atom and four O atoms.

### Key term

**hydroxide**: compound that contains one atom each of oxygen and hydrogen bonded together; for example, potassium hydroxide (KOH).

**3** Write the name of each of these compounds. Use the Periodic Table to help you find the name of the element from its symbol.

**a)** LiCl

**b)** CaO

**c)** MgS

**d)** $NaNO_3$

**e)** $K_2CO_3$

## Activity 6.1: Reacting sodium with water

A teacher showed his class the reaction of sodium with water.

He filled a large glass container with water and put in some Universal Indicator. He put some plastic screens between the container and the class.

He put on his splashproof eye protection and used a sharp knife to cut off a very small bit of sodium from a larger chunk.

6.4

He picked up the small bit using tweezers and added it to the water.

As soon as the sodium hit the water it gave off a gas which pushed it around the surface of the water. The reaction also gave out heat. The water turned purple.

**A1** The products of the reaction are hydrogen and sodium hydroxide. Write a word equation for the reaction.

**A2** Sodium hydroxide dissolves in water and is an alkali. What evidence is there that sodium hydroxide was made?

**A3** List the risks of this experiment.

**A4** Explain how the teacher controlled the risks to himself and the class.

## Rearranging atoms

During a chemical reaction the atoms in the reactants rearrange to form the products.

No atoms are lost. No new atoms are made.

*6.5 Some chemical reactions result in a precipitate being formed.*

Figure 6.5 shows a chemical reaction: colourless silver nitrate solution is reacting with colourless sodium chloride solution. One product formed is white silver chloride, which is insoluble. It is a white solid and sinks to the bottom of the beaker. The silver chloride is a precipitate. The word equation is:

| | | | | | | |
|---|---|---|---|---|---|---|
| silver nitrate | + | sodium chloride | → | silver chloride | + | sodium nitrate |

Chemical reactions can also be shown as particle diagrams. They show you what happens to the atoms. The particle diagram for this reaction is:

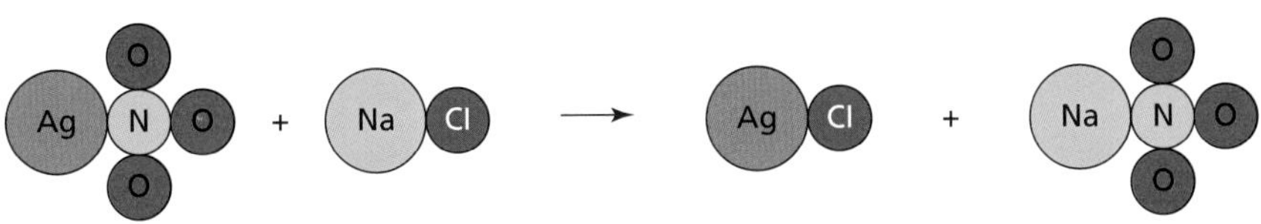


**6.6** *A particle diagram to show the reaction between silver nitrate and sodium chloride.*

You can see that there are seven atoms on each side of the equation. No atoms have been lost or gained during the reaction.

The atoms of the elements in the reactants are also present in the products.

4 Carbon (C) reacts with oxygen ($O_2$) to produce carbon dioxide ($CO_2$). Show this reaction as a particle diagram.

## Science in context: Removing mercury from water

Mercury is released into the environment through processes such as burning fossil fuels and mining. It can form soluble mercury compounds and dissolve in water. Mercury is very toxic and can cause brain damage so it is important that it is removed from drinking water. This can be done by adding aluminium sulfate to the water. Aluminium sulfate reacts with the soluble mercury compounds to form an insoluble mercury compound that can be filtered out of the water.

**6.7** *Polluted water containing mercury can enter rivers from factories and mines.*

## Key facts:

- ✔ During a chemical reaction, reactants react to form new products.
- ✔ A word equation separates the substances that react (on the left) and the products that are formed (on the right) with an arrow.
- ✔ Some substances are generally unreactive and can be described as inert.

## Check your skills progress:

- I can use symbols to represent chemical formulae.
- I can make risk assessments for practical work to identify and control risks.

Chapter 6 . Topic 2

# Pure substances and mixtures

You will learn:

- To describe correctly the purity of a mixture
- To know that reactions can produce single pure products or impure mixtures
- To identify and control risks for practical work
- To discuss how scientific knowledge is developed
- To describe the application of science in society, industry and research
- To describe how scientific progress is made through individuals and collaboration
- To discuss the global environmental impact of science

## Starting point

| You should know that... | You should be able to... |
| --- | --- |
| A mixture is made up of at least two different elements or compounds | Identify risks in a practical activity |
| New products are made during a chemical reaction | |
| An insoluble solid can be separated from a liquid in a mixture | |

## Pure substances

A solution is a mixture of a solute and a solvent.

The solute is soluble, it dissolves into the solvent, forming the solution.

The label on a bottle of mineral water might claim that it is 'pure water'. However, mineral water is a **mixture** – it is water with other chemicals dissolved in it. A **pure** substance only contains one type of particle.

**6.8** *Mineral water is a mixture.*

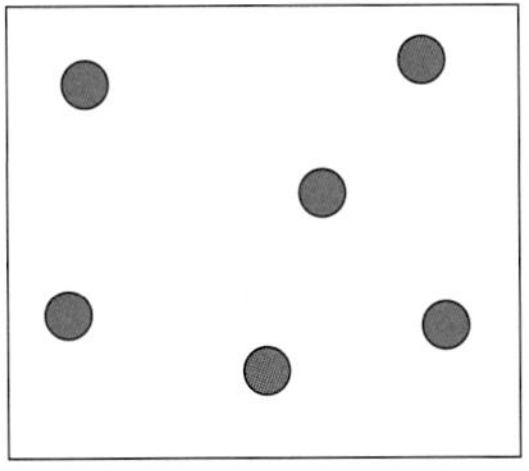

pure substance

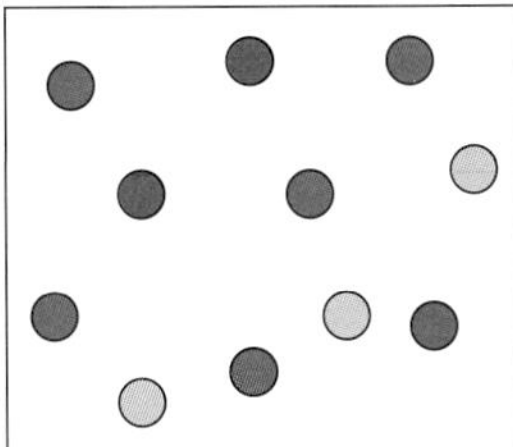

mixture

**6.9** *A pure substance contains only one type of particle, a mixture contains many different types.*

### Key terms

**mixture**: two or more elements or compounds mixed together. They can easily be separated.

**pure**: substance that contains only one element or compound.

Pure substances can be an element or a compound.

For example, pure gold is an element – it just contains gold particles.

Pure water is a compound – it only contains water particles.

1. Air is a mixture. Explain what this means.
2. Jofin claims that carbon dioxide is a mixture because it contains both carbon and oxygen atoms. Is he correct? Explain why.

### Key terms

**alloy**: a mixture of metal with other elements.

**purity**: how much of a chemical is in a mixture.

## Purity

The **purity** of a mixture tells us how much of a chemical is in a mixture. Water with high purity has a low amount of other chemicals mixed with it.

Pure metals are often mixed with other metals to form a mixture called an **alloy**. The purity of the alloy shows us how much of the original metal is in the mixture.

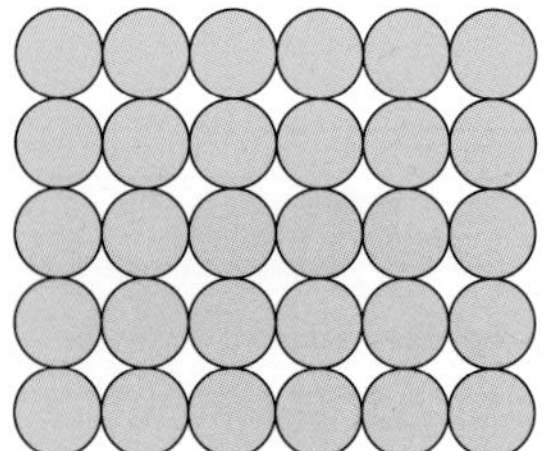


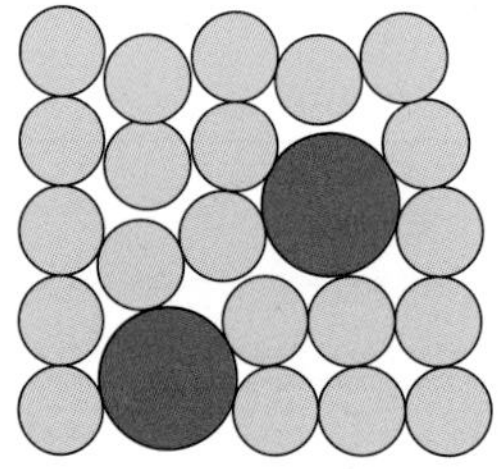


**6.10** *Pure metals only contain one type of atom. Alloys are a mixture, so they contain different types of atom.*

**6.11** *The purity of gold is measured in karats – the higher the karats the more pure the gold is. Pure gold is 24 karat.*

Adding other chemicals to a substance affects its properties. For example, pure water melts at 0 °C and boils at 100 °C. If you add other substances to the water, and change its purity, these temperatures change. For example, seawater boils at around 104 °C.

3. Yasmin cooled down some water. It froze at –5 °C. What does this tell her about the purity of the water?
4. 18 karat gold is 75% pure gold. An 18 karat gold ring has a mass of 4 g. How much gold does it contain?

## Products from a chemical reaction

Scientists use chemical reactions to make useful products.

Reactions do not always result in a single pure product. Often, they produce a mixture of products.

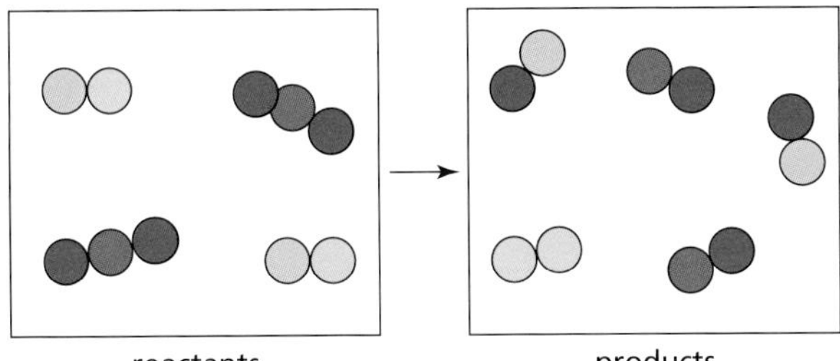


**6.12** *In this reaction the products are a mixture of different compounds.*

**Neutralisation** reactions are used to make **salts**.

They can be shown as a word equation:

Acid + alkali → salt + water

The two products are salt and water. To get a pure salt, you have to separate it from the water.

## Risk assessments

Doing chemical reactions involves risks. In order to identify and control risks scientists write **risk assessments**.

Hazards are things that can cause harm. You might see **hazard** labels on the substances you use.

A **risk** is the chance of a hazard causing harm to you, or people around you.

When you write a risk assessment you think about the risks from each hazard, and how to control them.

The first step in a reaction might be to pour some acid into a test tube. A suitable risk assessment is shown in table 6.2.

| Hazard | Risk | Controlling the risks |
|---|---|---|
| Corrosive acid | Can cause burns | Wear eye protection<br>Put the test tube into a test tube rack before pouring in the acid. Do not hold it |
| Glass test tube | The test tube could break and cause cuts if touched | Tell the teacher if anything breaks. Do not pick up any broken glass yourself |

**Table 6.2** *Writing a risk assessment.*

### Key terms

**hazard**: harm that something may cause.

**neutralisation**: chemical reaction between an acid and an alkali, which produces a neutral solution

**risk**: chance of a hazard causing harm.

**risk assessment**: the process of identifying the hazards involved in a practical investigation and deciding how to control them.

**salt**: a type of compound that consists of metal atoms joined to non-metal atoms, e.g. sodium chloride.

## Activity 6.2: Investigating purity

A scientist decides to use a neutralisation reaction to make copper sulfate.

The reaction produces a mixture of copper sulfate dissolved in water.

**A1** Plan a method she could use to get pure copper sulfate from the mixture.

**A2** Write a risk assessment for this investigation.

**A3** Explain how she could test the purity of the copper sulfate she collects.

**6.13** *A beaker of copper sulfate solution.*

**5** Evie uses a reaction to produce an insoluble substance and water. Describe how she could separate the two products.

## Science in context: Reducing waste products

Chemical reactions are used to make many useful products, such as medicines and plastics. Often only one of the products formed is useful, the others being waste products. Waste products are sometimes harmful to people or the environment, but removing and disposing of them costs money and uses energy. So, scientists work together, researching new reactions that form purer products.

**6.14** *Scientists research reactions that produce pure substances.*

## Key facts:

- ✔ A pure substance only contains one type of particle.
- ✔ The purity of a mixture tells us how much of a specific chemical is in a mixture.
- ✔ Reactions do not always result in a single pure product. Often, they produce an impure mixture of products.

## Check your skills progress:

- I can make risk assessments for practical work to identify and control risks.

Chapter 6 . Topic 3

# Measuring temperature changes

You will learn:

- To use temperature change to identify chemical reactions
- To make predictions based on scientific knowledge and understanding
- To plan investigations, including fair tests, while considering variables
- To identify and control risks for practical work
- To choose experimental equipment and use it correctly
- To carry out practical work safely
- To evaluate the reliability of measurements and observations
- To present and interpret scientific enquiries correctly
- To take accurate measurements and explain why this matters
- To use results to describe the accuracy of predictions
- To reach conclusions studying results and explain their limitations

## Starting point

| You should know that... | You should be able to... |
|---|---|
| During a chemical reaction the reactants react to form new products | Plan investigations by considering important variables, choosing which to change or keep the same |
| A temperature change can indicate that a chemical reaction has taken place | Present measurements appropriately |
| | Evaluate investigations |

## Observing temperature changes in reactions

There are a number of signs which show that a chemical reaction is taking place. One of them is a change in temperature. In some chemical reactions, energy is transferred to the surroundings as heat, which causes the temperature of the surroundings to increase. In other chemical reactions heat energy must be transferred *to* the reaction to make it happen.

**6.15** *Burning wood is a chemical reaction. Energy is released to the surroundings as heat.*

1 Look at the data in the table. For each reaction, state whether energy has been transferred to or from the reacting substances.

| Reaction | Temperature at start (°C) | Temperature at end (°C) |
|---|---|---|
| **a** magnesium ribbon with dilute sulfuric acid | 20 | 31 |
| **b** copper(II) sulfate solution with magnesium powder | 19 | 57 |
| **c** sodium hydrogen carbonate with citric acid | 20 | 17 |
| **d** sodium hydroxide with dilute hydrochloric acid | 19 | 21 |

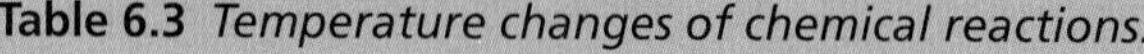

**Table 6.3** *Temperature changes of chemical reactions.*

**6.16** *Chemical reactions in a hand warmer transfer energy to the surroundings as heat.*

The temperature changes that happen during chemical reactions can be very useful.

Figure 6.16 shows a chemical hand warmer. The pack contains iron powder and water. When you open the pack, the iron powder and water react with oxygen from the air. The reaction transfers energy to the surroundings as heat. This causes the hand warmer to heat up.

Some instant cold packs used to treat sports injuries also use chemical reactions. When baking soda (sodium hydrogen carbonate) is added to citric acid solution they react to form sodium citrate, carbon dioxide and water. The pack gets cold because energy is transferred to the reaction mixture from the surrounding pack and this causes the temperature of the pack to decrease.

## Activity 6.3: Which has the biggest temperature change?

Plan an investigation to determine which of the following solutions produces the biggest temperature change in their reaction with a dilute acid:

- sodium hydroxide
- potassium hydroxide
- sodium hydrogen carbonate
- potassium hydrogen carbonate

**A1** Make a risk assessment to identify and control risks.

**A2** Write an apparatus list, name the variables and describe the method for this investigation – think carefully about what you will measure, what variable or variables you will keep the same to ensure a fair test, and how many measurements you should make.

**A3** Describe how you will take **accurate** and **precise** measurements and why this is important.

**A4** Draw a results table for this investigation.

## Evaluating investigations

The method you planned in Activity 6.3 should allow you to collect results to help you to answer the question and to be confident in your results. However, you may have realised that the hot reaction mixture transfers energy to the surroundings. This affects the temperatures you measure – **evaluating an investigation** means describing:

- what you would do differently if you did it again
- why you would do these things differently.

You should consider:

- Are there enough results to support the conclusion? Are there any repeat measurements?
- Is the method suitable?
- Are all the variables controlled?
- Are there any sources of error?
- Were measurements taken accurately and precisely?

**2** Evaluate the investigation you planned in Activity 6.3. How could you improve your method?

### Key terms

**accurate**: an accurate result is one that is close to the real answer.

**evaluating an investigation**: describing what you would do differently in an experiment if you repeated it and explaining why you would do those things differently.

**precise**: how precise your measurement is depends on the measuring equipment and the smallest difference it can measure. The smaller the difference it can measure, the more precise the measuring equipment is.

## Temperature changes during dissolving

Energy transfers also occur during the process of dissolving. This is a physical change, not a chemical change. When some salts, such as potassium chloride and ammonium nitrate, are dissolved in water the temperature decreases. When other salts, such as calcium chloride, are dissolved in water the temperature increases.

One type of instant cold pack (figure 6.17) uses dissolving to make the pack cold. Squeezing the pack bursts an inner bag so that water sealed inside mixes with powdered ammonium nitrate. When the ammonium nitrate dissolves energy is transferred to the ammonium nitrate from the water surrounding it. The temperature of the water decreases, making the pack cold.

**6.17** *An instant cold pack that can be used to treat an injured leg.*

## Activity 6.4: Investigating temperature changes during dissolving

Some students wanted to investigate how changing the mass of ammonium chloride affected the temperature change when ammonium chloride dissolves in water.

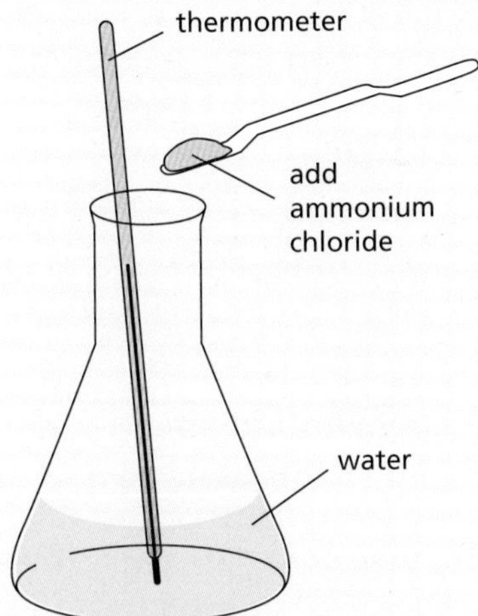


**6.18** *Measuring the temperature change when ammonium chloride is added to water.*

**A1** What is the independent variable for this investigation?

**A2** What is the dependent variable for this investigation?

**A3** Name one control variable for this investigation.

**A4** Predict the results of this investigation – what effect will an increase in the mass of ammonium chloride have on the temperature? Give a reason.

**A5** **a** What column headings should be used for the results table? Include the units.

**b** How should the results be presented – as a chart or graph?

### Making links

You may have previously learned in physics that when energy is transferred from one object to another some energy is transferred to the environment. This can be in the form of heat, as we see in these chemical reactions. What other ways can energy be transferred to the surroundings in a chemical reaction?

### Key facts:

✔ During chemical reactions, energy can be transferred to or from the surroundings.

✔ The transfer of thermal (heat) energy to or from the surroundings causes the temperature to change.

### Check your skills progress:

- I can plan an investigation using previous knowledge and understanding.
- I can evaluate an investigation, suggesting improvements and explaining any proposed changes.
- I can make predictions of likely outcomes based on scientific knowledge.
- I can make risk assessments for practical work to identify and control risks.
- I can explain why accuracy and precision are important.

Chapter 6 . Topic 4

# Exothermic and endothermic processes

You will learn:

- To use temperature change to identify endothermic and exothermic reactions
- To carry out practical work safely
- To collect and record observations and measurements appropriately
- To present and interpret scientific enquiries correctly
- To describe results in terms of any trends and patterns and identify any abnormal results
- To reach conclusions studying results and explain their limitations
- To describe the application of science in society, industry and research
- To discuss the global environmental impact of science

## Starting point

| You should know that... | You should be able to... |
|---|---|
| A temperature change can indicate that a chemical reaction has taken place | Identify variables within an investigation |
| Change of state is a physical process | |
| During evaporation some of the particles in the liquid leave as a gas | |

## Energy changes in physical processes

Melting and freezing are examples of physical processes. An ice cube will begin to melt if you take it out of the freezer. It will change from the solid state to the liquid state.

This process happens because **thermal energy** is transferred from the warm room to the cold ice, overcoming forces between the water particles. Melting is an **endothermic** process because energy is transferred from the surroundings while it happens.

A tray of water from the tap will freeze if you put it in a freezer. It will change from the liquid state to the solid state.

This process happens because thermal energy is transferred from the warm water to the cold freezer, allowing forces to form between the water particles. Freezing is an **exothermic** process because energy is transferred to the surroundings while it happens.

### Key terms

**endothermic**: reaction or process in which energy is transferred from the surroundings, usually by heating, causing the temperature of the surroundings to decrease.

**exothermic**: reaction or process in which energy is transferred to the surroundings, usually by heating, causing the temperature of the surroundings to increase.

**thermal energy**: energy stored in an object due to its temperature.

Is boiling an exothermic process or an endothermic process? Explain your answer.

Exothermic and endothermic processes often cause a change in temperature. If you hold some ice, your hands become cold as thermal energy is transferred from them to the melting ice. Endothermic processes cause a decrease in temperature of the surroundings.

**6.19** *Ice melting is an endothermic process.*

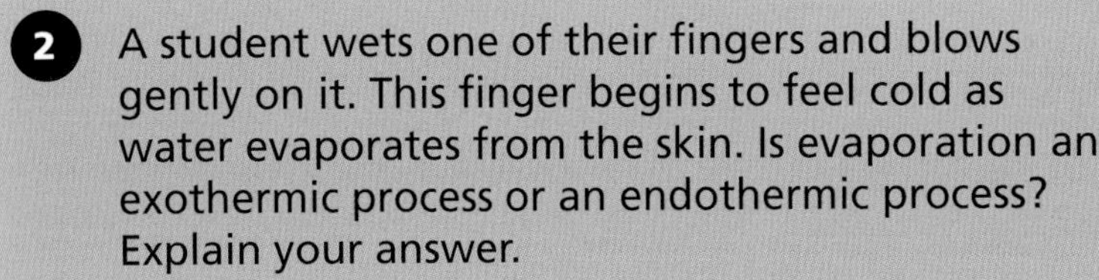

**2** A student wets one of their fingers and blows gently on it. This finger begins to feel cold as water evaporates from the skin. Is evaporation an exothermic process or an endothermic process? Explain your answer.

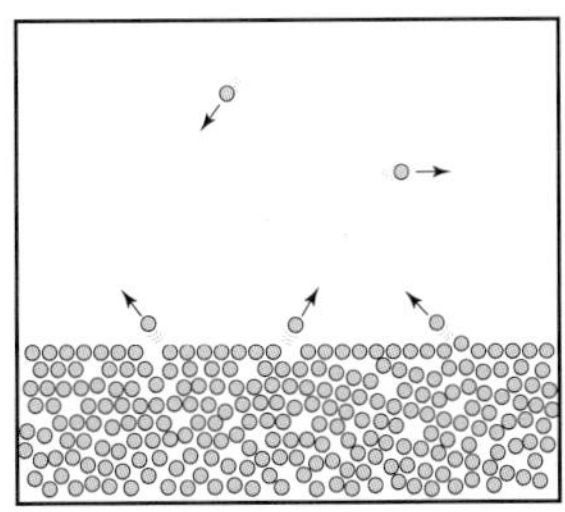

**6.20** *During evaporation, some of the particles in the liquid overcome all the forces between them and leave as a gas.*

Dissolving a substance in water can be an exothermic process or an endothermic process, depending on the substance being dissolved. For example, if you dissolve calcium chloride in water, the mixture becomes hot. Exothermic processes cause an increase in temperature.

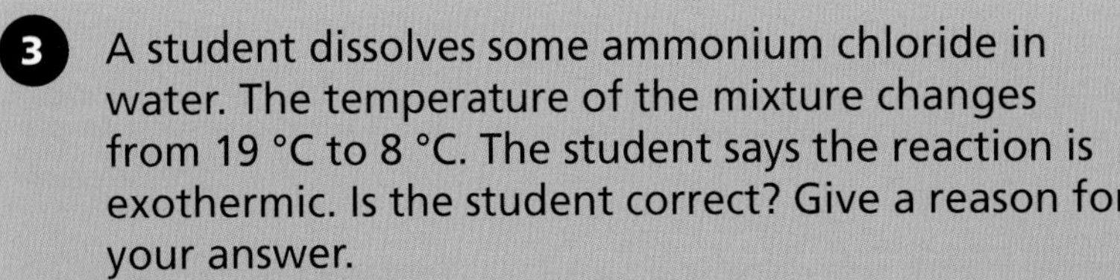

**3** A student dissolves some ammonium chloride in water. The temperature of the mixture changes from 19 °C to 8 °C. The student says the reaction is exothermic. Is the student correct? Give a reason for your answer.

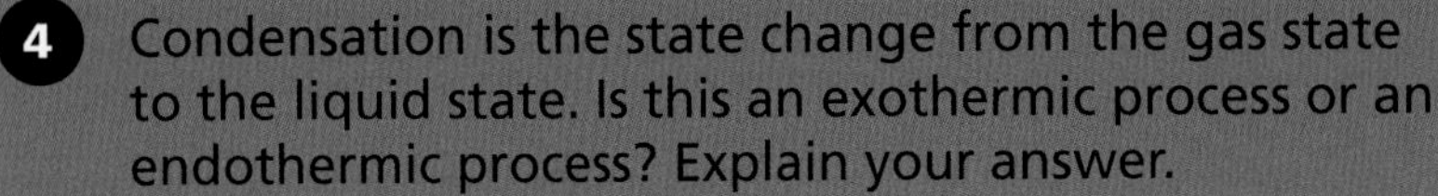

**4** Condensation is the state change from the gas state to the liquid state. Is this an exothermic process or an endothermic process? Explain your answer.

**6.21** *Cold packs can ease the pain of a sports injury. Some contain ammonium chloride and water, which are mixed together when the cold pack is needed.*

## Energy changes in chemical reactions

All **chemical reactions** involve energy transfers. It is obvious in many exothermic reactions that energy is being transferred to the surroundings. When a fuel burns, energy is transferred to the surroundings by light and by heating. You can see the bright flame and feel its heat. Energy may also be transferred to the surroundings by sound. You can hear the bangs from fireworks exploding at a fireworks display.

### Key term

**chemical reaction**: a change in which new substances are produced.

**6.22** *A fuel burning is an exothermic reaction.*

**6.23** *Fireworks use exothermic reactions to produce loud explosions. The colours in fireworks come from the light that is also emitted.*

Common examples of exothermic reactions include:

- Combustion – fuels and other substances burning.
- Reactions between metals and acids
- Calcium oxide reacting with cold water – this is so exothermic that the water boils after a few minutes.
- Neutralisation reactions between acids and alkalis

**5** When a person is exercising, their muscles contract frequently and their rate of respiration increases. The process of respiration involves several reactions. Describe how you can tell that the reactions involved in respiration are exothermic overall.

**6** Explain why some exothermic reactions may be dangerous to do in a school laboratory.

## Science in context: The thermit reaction

The thermit reaction involves aluminium and iron oxide:

aluminium + iron oxide → aluminium oxide + iron

The thermit reaction is very exothermic. Iron is one of the products. It melts during the reaction because so much energy is transferred to it by heating. The thermit reaction is useful because the molten iron can be used to join railway track together.

**6.24** *This railway worker is using the thermit reaction. Molten iron runs into the gap between two lengths of railway track, joining them together when it cools and solidifies.*

## Science in context: Photosynthesis

Plants use sunlight for photosynthesis. This is an important endothermic reaction for living things. Energy must be transferred to the plant for photosynthesis to happen. Unlike most other endothermic reactions, the energy needed for photosynthesis is transferred from the surroundings by light.

**6.25** *Photosynthesis is an endothermic reaction in plants that needs sunlight.*

## Endothermic reactions

Most chemical reactions are exothermic but some are endothermic. A **precipitate** of calcium carbonate forms when sodium carbonate solution reacts with calcium chloride solution:

sodium carbonate + calcium chloride →
calcium carbonate + sodium chloride

The temperature of this reaction mixture decreases, showing that it is an endothermic reaction. Note that most other precipitation reactions are exothermic reactions.

### Key term

**precipitate**: insoluble solid formed when soluble substances react together.

However, most endothermic reactions are difficult to identify by looking for a temperature change. This is because energy must be transferred continuously to the reactants for these reactions to happen.

When green copper carbonate powder is heated, it breaks down to form black copper oxide powder and carbon dioxide gas. This reaction is easy to see because of the colour change and the movement of the powder caused by the escaping gas. However, the reaction stops as soon as the heating stops, showing that it is an endothermic reaction.

7 Explain why photosynthesis does not happen in the dark.

8 Draw a table that summarises the differences between exothermic and endothermic changes. For each type of change, include at least *two* examples of a physical process and of a chemical reaction.

9 When an electric current is passed through liquid calcium chloride, the liquid breaks down to form calcium and chlorine. This reaction stops happening when the electricity supply is turned off. Is this an exothermic chemical reaction or an endothermic chemical reaction? Explain your answer.

### Making links

Precipitation reactions are covered in Stage 7 Topic 5.2. Suggest **three** ways that you can tell a precipitation reaction has happened.

### Making links

You will study photosynthesis in Stage 9 Topic 1.1. Plants can carry out this chemical reaction but animals cannot. In the process of photosynthesis, plants take in light energy and use it to convert carbon dioxide and water into chemical energy, glucose and oxygen. Suggest a reason why photosynthesis is important for animals and other living things, not just plants.

## Science in context: Fuel cell cars

The reaction between hydrogen and oxygen is exothermic. When this reaction happens in a hydrogen-oxygen fuel cell, energy is transferred to the surroundings as an electric current in a circuit. In a fuel cell car, this used to run an electric motor connected to the wheels.

Water is the only product emitted by a hydrogen-oxygen fuel cell car when it is in use. This is different from petrol and diesel cars, which emit carbon dioxide. This is a greenhouse gas and a major cause of global warming. The use of hydrogen-oxygen fuel cells could be a useful way to reduce carbon dioxide emissions. It could also reduce the world's reliance on crude oil, which is a non-renewable resource that will run out if we keep using it.

**6.26** *Some electric cars use hydrogen-oxygen fuel cells instead of rechargeable batteries.*

10 Hydrogen for hydrogen-powered cars can be made by passing an electric current through water. Oxygen is also produced in the reaction. Identify the type of reaction involved, and explain your answer.

## Activity 6.5: Comparing alcohol fuels

Some students completed an experiment to compare the energy transferred when different alcohol fuels burn. Figure 6.27 shows the apparatus they used.

The students followed this method.

**A** Weigh a spirit burner of fuel on a digital balance and record the mass.

**B** Add 200 $cm^3$ of water to the container and record its initial temperature.

**C** Put the spirit burner under the container, then light the wick.

**D** Heat the water until its temperature increases by 10 °C. Extinguish the flame.

**E** Reweigh the spirit burner to find the mass of the fuel burned.

spirit burner

**6.27** *Measuring the temperature rise produced by burning a liquid fuel.*

The table shows the students' results.

| Alcohol fuel | Initial mass (g) | Final mass (g) |
|---|---|---|
| Ethanol | 80.54 | 79.13 |
| Methanol | 87.52 | 85.68 |
| Propanol | 76.14 | 74.90 |

**Table 6.4** *Changes in mass when alcohol fuels are burned.*

**A1** State the independent variable, dependent variable, and at least one control variable in this experiment.

**A2** Calculate the mass of each fuel needed to increase the temperature of the water by 10 °C. Use the information in the table.

**A3** Which fuel produces the most energy per gram burned?

**A4** Write a conclusion for the experiment.

**A5** Evaluate the method used and identify any improvements you would make.

### Key facts:

- ✔ Energy is transferred to the surroundings by exothermic processes and chemical reactions, causing the temperature of the surroundings to increase.
- ✔ Energy is transferred from the surroundings by endothermic processes and chemical reactions, causing the temperature of the surroundings to decrease.

### Check your skills progress:

- I can make conclusions by interpreting results.
- I can evaluate experiments, suggesting improvements, explaining any proposed changes.

Chapter 6 . Topic 5

# The reactivity series

You will learn:

- To describe the reactivity of metals with oxygen, water and dilute acids
- To make predictions based on scientific knowledge and understanding
- To plan investigations, including fair tests, while considering variables
- To identify and control risks for practical work
- To collect and record observations and measurements appropriately
- To use results to describe the accuracy of predictions
- To reach conclusions studying results and explain their limitations
- To describe the application of science in society, industry and research

## Starting point

| You should know that... | You should be able to... |
|---|---|
| Evidence that a chemical reaction has happened includes the formation of a gas, formation of a precipitate, a change in colour and a change in temperature | Plan investigations to test ideas |
| During a chemical reaction, reactants interact to form products | Select and use a range of apparatus correctly |
| Hydrogen gas can be identified using a simple laboratory test | Make predictions using scientific knowledge and understanding |

## Reactions and reactivity

Some substances do not easily take part in chemical reactions, and may not react with some other substances at all. They have very low **reactivity** and can be described as being inert. Helium is so inert that chemists have been unable to get it to react with any other substances at all.

Some substances have very high reactivity and can be described as being very reactive. Fluorine is a very reactive gas. It can react with almost every other substance apart from helium.

### Key terms

**reactivity**: how likely it is that a substance will undergo a chemical reaction.

**reactivity series**: series of metals written in order from the most reactive to the least reactive.

Just as non-metals like helium and fluorine vary in their reactivity, metals vary in their reactivity too. Figure 6.29 shows some common metals listed in order of their reactivity, from the most reactive to the least reactive. A list like this is called a **reactivity series**.

In this reactivity series, potassium is the most reactive metal and gold is the least reactive metal. Hydrogen is a non-metal, but it is shown in this reactivity series so that you can make predictions about how metals will react with water and dilute acids.

**6.28** *The helium in these party balloons is inert and cannot be set on fire.*

most reactive
potassium
sodium
calcium
magnesium
zinc
iron
(hydrogen)
copper
silver
gold
least reactive

**6.29** *A reactivity series of metals, with hydrogen (a non-metal) included.*

1. Identify *three* metals that are less reactive than hydrogen.
2. Identify a metal that is more reactive than zinc but less reactive than calcium.

## Reactions of metals with water

When a metal reacts with cold water, the products are a metal hydroxide and hydrogen. This is a general word equation for these reactions:

metal + water → metal hydroxide + hydrogen

You can tell that one of these reactions is happening because bubbles of hydrogen gas are given off. No bubbles are given off if the metal does not react with cold water. The more reactive a metal is, the faster the rate of bubbling.

Potassium is the most reactive metal in the reactivity series. It reacts very vigorously with water:

potassium + water → potassium hydroxide + hydrogen

So much energy is transferred by heating that the hydrogen catches fire. Depending on the size of the piece of potassium, the reaction may only last a few seconds before the hot metal explodes with a small cracking sound. The potassium hydroxide dissolves in the water to make an **alkaline** solution. You can detect this using **Universal Indicator**, which turns purple.

### Key terms

**alkaline**: having the properties of an alkali.

**Universal Indicator**: type of indicator which can change into a range of colours depending on whether the solution is acidic or alkaline and how strong it is.

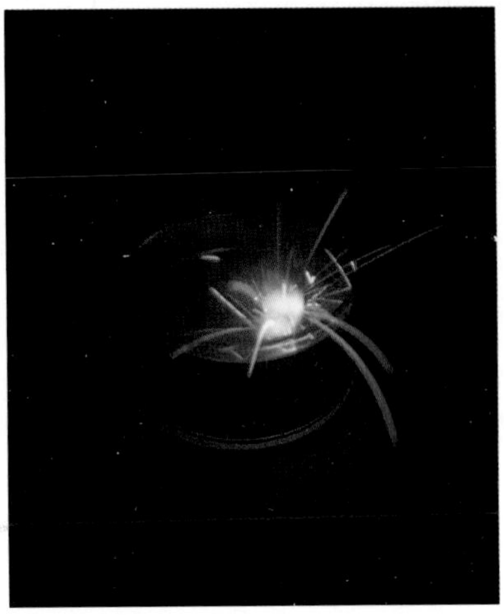

**6.30** *You see a lilac flame and sparks when potassium reacts with water.*

Calcium reacts steadily with cold water. However, the reaction is exothermic and so the reaction mixture quickly warms up. The bubbling becomes very quick but the hydrogen does not catch fire. Magnesium reacts very slowly with cold water, producing a few bubbles over a few days, and zinc does not react at all. Metals that are less reactive than hydrogen will not react with water.

3. Identify *three* metals, other than zinc, that will not react with water. Give a reason for your answer.
4. Explain why some roofs are made of zinc.
5. Predict how sodium will react with cold water. Include a word equation in your answer.

## Reactions of metals with dilute acids

When a metal reacts with a dilute acid, the products are a salt and hydrogen:

metal + acid → salt + hydrogen

The salt made depends on the reacting metal and acid. Sulfate salts are made when sulfuric acid is used, and chloride salts are made when hydrochloric acid is used. For example, magnesium reacts with dilute hydrochloric acid to make magnesium chloride:

magnesium + hydrochloric acid → magnesium chloride + hydrogen

You can also model this reaction using a particle diagram.

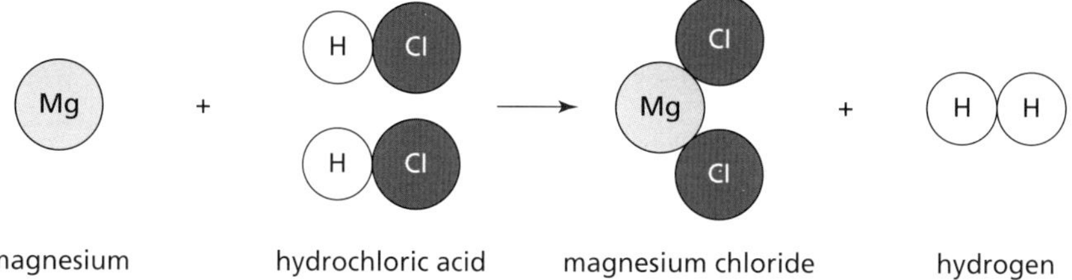


**6.31** *This particle diagram shows how magnesium reacts with hydrochloric acid.*

You can show that the bubbles are caused by hydrogen using a simple test. If you hold a lighted wooden splint in the gas, you will hear a 'pop' sound if it is hydrogen.

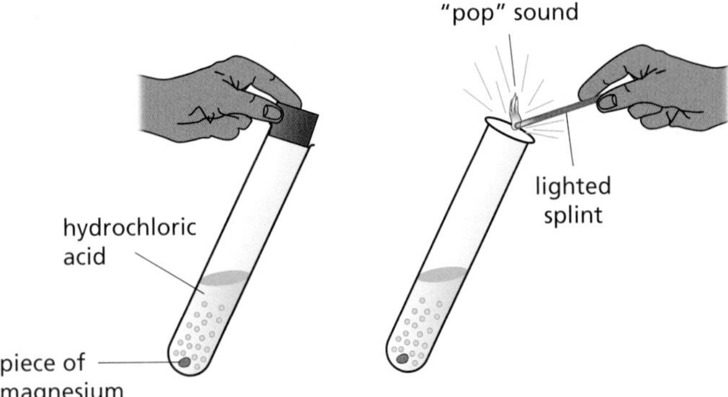


**6.32** *How to test for hydrogen.*

**6.33** *The reactions of four metals with dilute hydrochloric acid. Left to right: calcium, magnesium, zinc and copper.*

### Making links

You may have studied the laboratory tests for three gases, including the test for hydrogen, in Stage 7 Topic 6.5. Describe the laboratory tests for oxygen and carbon dioxide.

Metals will only react with dilute acids if they are more reactive than hydrogen. Potassium and sodium react very vigorously with dilute acids, and these reactions cannot be done safely in the school laboratory. Calcium reacts vigorously, magnesium rapidly, zinc slowly, and copper does not react at all.

**6** Name *two* metals, other than copper, that will not react with dilute acids. Give a reason for your answer.

**7** Predict how iron will react with dilute sulfuric acid. Include a word equation in your answer.

## Reactions of metals with oxygen

One product is formed when a metal reacts with oxygen in the air:

metal + oxygen → metal oxide

At room temperature, the surface of a reactive metal changes colour and becomes dull as a layer of its oxide forms.

Potassium and sodium react very quickly with oxygen. When they are cut with a knife, the cut surface is shiny and silvery. This changes to dull grey as the metal reacts with oxygen in the air. The change with potassium happens in seconds and the change with sodium happens in minutes:

sodium + oxygen → sodium oxide

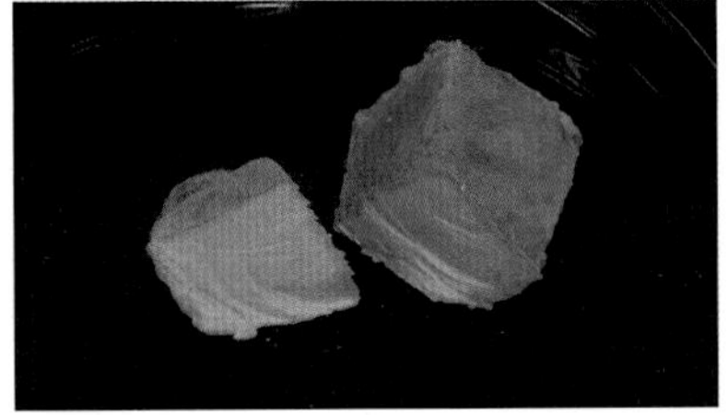

**6.34** *The shiny cut surface of sodium rapidly turns dull in air.*

A metal oxide layer protects calcium, magnesium and zinc from reacting with oxygen in air, unless these metals are heated in a Bunsen burner flame. Calcium burns rapidly with a red flame and magnesium burns quickly with a white flame.

Less reactive metals such as iron and copper do not burn when heated. Instead, iron forms a dull black layer of iron oxide, and copper forms a dull black layer of copper oxide. Silver and gold do not react with oxygen at all.

copper + oxygen ⟶ copper oxide

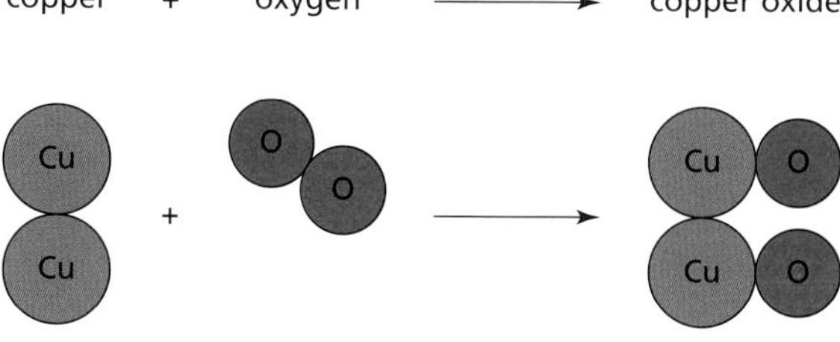


**6.35** *A word equation and particle diagram to model the reaction between copper and oxygen.*

**6.36** *When water or moisture are present, iron reacts slowly with oxygen in the air to form orange-brown rust.*

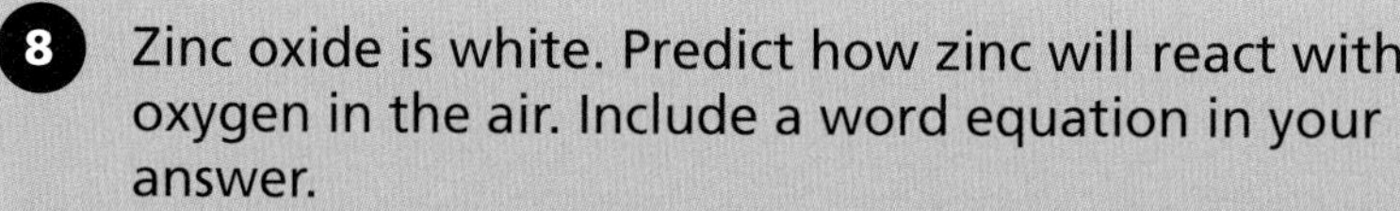

**8** Zinc oxide is white. Predict how zinc will react with oxygen in the air. Include a word equation in your answer.

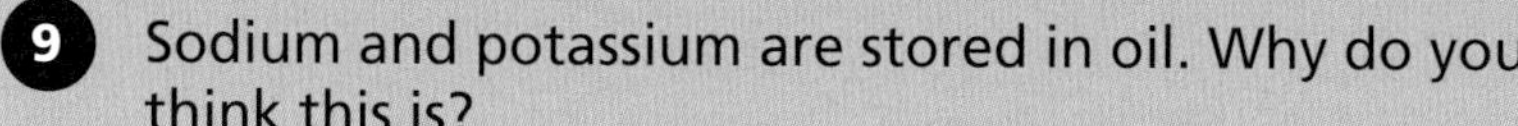

**9** Sodium and potassium are stored in oil. Why do you think this is?

## Science in context: Cells and batteries

Batteries and cells rely on the reactivity of metals such as zinc. Remember that a battery is made up of more than one single electrical cell joined together. In the familiar cylinder-shaped cells zinc gradually turns to zinc oxide as it reacts with a manganese compound. In some types of flat 'button' batteries, zinc gradually turns to zinc oxide as it reacts with silver oxide. When no more zinc can react, the cell 'goes flat' and no longer works.

Lithium is a metal that is less reactive than sodium but more reactive than calcium. There are several types of lithium battery but all can store a lot of energy. They are used where a lot of electricity is needed, such as in smartphones and laptops. Without rechargeable lithium batteries, devices like these would be a lot larger than they are today, and electric vehicles would not be able to travel very far at all before having to recharge.

**6.37** *The battery for this early mobile phone was so large and heavy, that the phone had a carry handle.*

## Activity 6.6: Comparing the reactivity of metals

You know that we can observe differences in the reactions of metals with a dilute acid. Use this to plan an investigation to compare the reactivity of some metals. You will use the results to put the metals in an order of reactivity, and then compare this to the reactivity series.

The metals you will be testing are:

- iron
- copper
- zinc
- magnesium

**A1** Make a prediction – which metal do you think will be the most reactive and which metal will be the least reactive? Why?

**A2** Write a list of the apparatus you will need for this experiment.

**A3** Write a step-by-step method explaining how to use the apparatus and collect the evidence that you need to answer the question. Think about what you will measure and/or observe, and also how many measurements or observations you will need to make.

**A4** Make risk assessments to identify any hazards in doing this experiment and describe how you will control the risks.

**A5** Draw a results table that you could use for this investigation.

**A6** If possible, carry out your planned investigation.

**A7** Write a conclusion. Say what the results show and compare this with your prediction. Describe the accuracy of your prediction, based on results, and suggest why they were or were not accurate.

**10** Some students completed an experiment to compare the reactivity of four metals (A, B, C and D) with dilute hydrochloric acid. The table shows their results:

| Metal | Observations |
|---|---|
| A | Some bubbling, could not hear the fizzing |
| B | No reaction |
| C | Vigorous reaction – rapid bubbling, could hear fizzing |
| D | Steady bubbling, could hear quiet fizzing |

**Table 6.5** *Observations of chemical reactions between different metals and hydrochloric acid.*

a) How did the students know that chemical reactions had taken place?

b) What gas was given off during the reaction?

c) Use the results to put the metals in order of reactivity, from most reactive to least reactive.

d) The students' teacher told them that if lead had been added, a few bubbles would form but only very slowly. She also told them that the reaction with aluminium would have been quite vigorous – there would have been quick bubbling and they would have heard some fizzing.

Use this information to rewrite the reactivity series in part (c) to include these two metals.

## Activity 6.7: Is the pattern of reactivity of metals with acids the same with oxygen and water?

Plan an investigation to help in answering this question. Use the ideas from Activity 6.6 to help you.

**A1** Write an apparatus list, all the variables and any hazards in doing this experiment. How will you control the risks?

**A2** Write a step-by-step method for your investigation. State how you are going to make sure that your results will be reliable.

## Key facts:

✔ Metals react in similar ways, but some are more reactive than others.

✔ Information about the reactivity of metals with oxygen, water and dilute acids is used to put them in the reactivity series.

## Check your skills progress:

- I can make predictions of likely outcomes, based on scientific understanding.
- I can plan a range of investigations, while considering variables appropriately and how many repetitions are needed to be reliable.
- I can make risk assessments for practical work to identify and control risks.
- I can decide what equipment is required to carry out an investigation.
- I can describe the accuracy of predictions, based on results and suggest why they were or were not accurate.
- I can make conclusions by interpreting results.

# End of chapter review

## Quick questions

1. Write a word equation for the reaction when zinc is heated in oxygen.
2. Describe the difference between a pure substance and a mixture.
3. What is purity?
4. Explain why some reactions produce an impure mixture of products.
5. When magnesium reacts with dilute hydrochloric acid in a test tube, the contents of the test tube get warm. Is this an endothermic or exothermic process?
6. When potassium chlorate dissolves in water, the temperature of the mixture decreases. Is this an endothermic or exothermic process?
7. Which *one* of the following statements is true?

   During an exothermic reaction:

   **(a)** Energy is transferred from the surroundings; the temperature of the surroundings decreases.

   **(b)** Energy is transferred to the surroundings; the temperature of the surroundings decreases.

   **(c)** Energy is transferred from the surroundings; the temperature of the surroundings increases.

   **(d)** Energy is transferred to the surroundings; the temperature of the surroundings increases.
8. Which *one* of these metals will react most quickly with cold water?

   **(a)** Calcium

   **(b)** Zinc

   **(c)** Copper

   **(d)** Iron

## Connect your understanding

9. Draw and label a particle model comparing pure gold with impure gold.

10. Instant cold packs can be made using ammonium nitrate and water. The pack contains a small bag of water. This bag is inside another bag, which contains solid ammonium nitrate. When you squeeze the outer bag the smaller bag bursts. The ammonium nitrate dissolves in the water and the pack becomes cold.

    **(a)** Is this an endothermic process or an exothermic process?

    **(b)** Describe how energy is transferred between the reaction mixture and its surroundings.

    **(c)** Explain how you know this.

11. Jonathan carried out an experiment to measure the temperature change when two solutions are mixed. His results are shown below:

| Reaction | Temperature at start (°C) | Temperature at end (°C) |
|---|---|---|
| sodium hydrogen carbonate solution + citric acid solution | 21 | 18 |
| sodium hydroxide solution + citric acid solution | 21 | 23 |
| dilute hydrochloric acid + citric acid solution | 21 | 21 |

    **(a)** Identify the dependent and the independent variables in this experiment.

    **(b)** State *one* variable that Jonathan should have controlled in this experiment.

    **(c)** Which *one* of these reactions would be the most suitable for use in making an instant cold pack? Explain your answer.

    **(d)** Based on Jonathan's results, what type of reaction is the second reaction?

12. Some students investigated the reactivity of three different metals (A, B and C) by adding each to a test tube containing dilute hydrochloric acid. The table shows their observations.

| Metal | Observations |
|---|---|
| Metal A | Lots of bubbles produced, fizzing heard, test tube became warm |
| Metal B | Lots of bubbles produced very quickly, test tube was hot to the touch |
| Metal C | Bubbles produced slowly, could hear fizzing if you listened very carefully, the temperature did not change |

    **(a)** Write a reactivity series for the metals in these reactions.

    **(b)** Explain your answer.

## Challenge questions

13. Lithium is less reactive than sodium but more reactive than magnesium.

(a) Lithium burns in air when it is heated. Write a word equation for this reaction.

(b) Lithium reacts quickly with water. Write a word equation for this reaction.

(c) Lithium reacts very quickly with dilute hydrochloric acid. Write a word equation for this reaction.

(d) Predict whether these reactions are exothermic or endothermic. Explain your answer.

14. Read this information about aluminium, then answer the questions.

Aluminium is almost as reactive as magnesium. A natural layer of aluminium oxide on a piece of aluminium prevents air and water reaching the metal below. Aluminium and aluminium oxide do not dissolve in water, but they do react with dilute hydrochloric acid. Soluble aluminium chloride is formed in these reactions. A piece of aluminium reacts with dilute hydrochloric slowly at first, but the reaction speeds up after a short time.

(a) Explain why a piece of aluminium does not react when placed in oxygen or water.

(b) Suggest what reaction first takes place when a piece of aluminium is placed in dilute hydrochloric acid. Give a reason for your answer.

(c) Explain why the reaction between a piece of aluminium and dilute hydrochloric acid speeds up after a short time.

# End of stage review

1. Scientists use models to show the structure of atoms.

   (a) Rutherford and his team carried out an investigation. They fired alpha particles at a thin sheet of gold foil. This diagram shows the pathways of the alpha particles.

   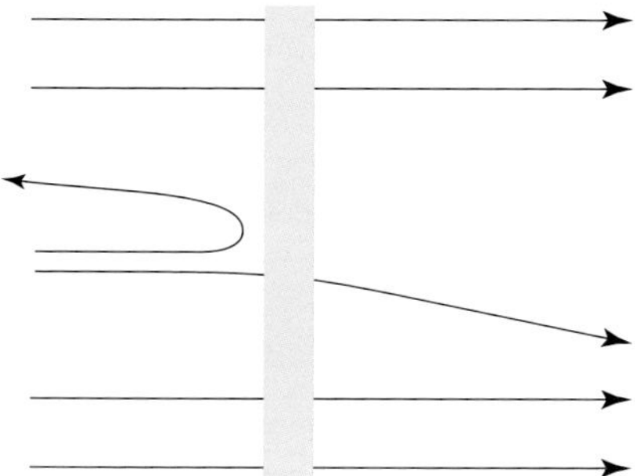

   i) Some particles were deflected. Name the part of the atom that caused this deflection.

   ii) Explain how the results showed that this part was very small in relation to the whole atom.

   (b) Explain why the model of the atom has changed over time.

2. Chromatography can be used to separate the dyes in ink. Evie uses chromatography on an ink. The chromatogram shows two spots. What conclusions can she make about the ink?

3. Some students investigated the reactions of four metals (A, B, C and D) with acid. They added equal masses of four different metals to 5 $cm^3$ of dilute hydrochloric acid and measured the temperature change. The table shows their results.

| Metal | Temperature change (°C) |
|---|---|
| A | 5.3 |
| B | 0.1 |
| C | 9.1 |
| D | 7.4 |

   (a) Identify the independent variable in this investigation.

   (b) Name this type of reaction.

   (c) Describe what happens during this reaction in terms of energy transfer.

**(d)** Name a metal that could give the results seen for metal B.

**(e)** Sort these four metals in order of increasing reactivity.

**(f)** Fizzing was seen in some of these reactions. Describe a laboratory test to confirm that the gas given off was hydrogen.

**4.** Kris added copper sulfate to 100 $cm^3$ of water until no more would dissolve. It formed a blue solution.

**(a)** Name the:

**i)** Solute

**ii)** Solvent

**(b)** The solution is not pure copper sulfate. Explain why.

**(c)** Describe how he could dissolve more copper sulfate in the same volume of water.

# Physics


## Chapter 7: Measuring motion

7.1: Measuring distance and time 133
7.2: Speed and average speed 136
7.3: Distance/time graphs 141
End of chapter review 145

## Chapter 8: Forces

8.1: Balanced and unbalanced forces 149
8.2: Turning effect of a force 155
8.3: Pressure on an area 161
8.4: Pressure and diffusion in gases and liquids 166
End of chapter review 175

## Chapter 9: Light

9.1: Reflection 179
9.2: Refraction 184
9.3: Coloured light 189
End of chapter review 196

## Chapter 10: Magnets

10.1: Magnets and magnetic materials 200
10.2: Electromagnets 208
End of chapter review 215
End of stage review 218

7

# Chapter 7

## Measuring motion

Being able to measure distance and time accurately is important in our world today. In sport, times of 1/100th of a second can make the difference between winning or losing a race. In space science, a spacecraft being a few centimetres out of position can mean it will miss its target and its mission will fail.

You will learn about:

- How distance and time are measured accurately
- How to calculate the speed of a moving object
- How to describe changes in the way an object moves
- How to use graphs to show how an object is moving

You will build your skills in:

- Using symbols and formulae to represent scientific ideas
- Making predictions using your knowledge and understanding of science
- Collecting and recording sufficient observations and/or measurements using an appropriate method
- Presenting observations and measurements appropriately

Chapter 7 . Topic 1

# Measuring distance and time

You will learn:

- To choose experimental equipment and use it correctly
- To take accurate measurements and explain why this matters
- To describe the application of science in society, industry and research

## Starting point

| You should know that... | You should be able to... |
|---|---|
| When things speed up, slow down or change direction there is a cause | Choose appropriate apparatus and use it correctly |
| Distance is measured in metres | Take accurate measurements |
| Time is measured in seconds | Know when you have repeated measurements enough times to be sure your results are reliable |

When you investigate how an object is moving you need to measure the distance travelled and the time taken. You need to think about how **accurate** your measurement needs to be and know the best equipment to use to get **precise** measurements.

### Key terms

**accurate**: an accurate result is one that is close to the real answer.

**precise**: how precise your measurement is depends on the measuring equipment and the smallest difference it can measure. The smaller the difference it can measure, the more precise the measuring equipment is.

**reliable**: measurements are reliable when repeated measurements give results that are very similar.

## Choosing the best apparatus to measure distance

Choose the best apparatus from the box to measure each of the following distances.

15 cm ruler marked in millimetres
metre rule marked in millimetres
10 m measuring tape marked in centimetres
100 m measuring tape marked in centimetres

**a)** How far you can run in 10 s.

**b)** The height of a table.

**c)** The length of your thumb.

**d)** How far you can walk in 2 s.

When you measure small distances, you need to be more precise than when you are measuring larger distances. For example, you would measure the width of your text book to the nearest millimetre, but the height of a wall to the nearest centimetre. Taking more precise measurements will make your results more accurate. Remember, you need to make sure your results are **reliable**.

## Activity 7.1: Measuring distance

Measure the following:

- height of your chair
- length of your pencil or pen
- width of your classroom
- length of your thumbnail.

**A1** What did you use to measure each distance?

**A2** For each measurement, write whether it is accurate to the nearest millimetre or centimetre.

**2** Why is it more accurate to measure the width of your desk with a metre rule than a 15 cm one?

## Choosing the best apparatus to measure time

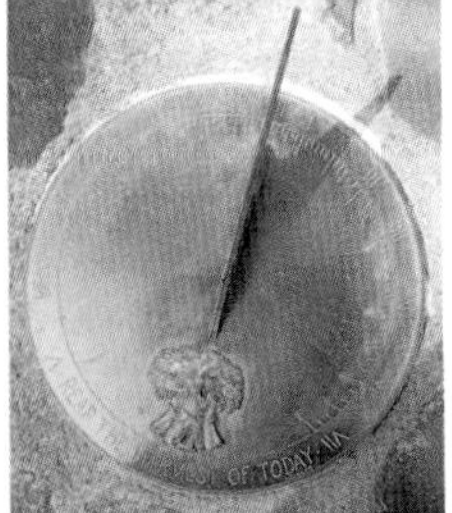

**7.1** *Different apparatus for measuring time.*

There are lots of ways to measure time (see figure 7.1). If you want to measure how long a lesson lasts, then you can use an ordinary clock or watch. If you need to measure the time to run 20 m you need more accurate equipment.

**3** Why is it important to repeat your measurements if you are using a stopwatch to time how long it takes a toy car to travel down a ramp?

## Activity 7.2: How long does it take?

Time how long it takes you to do the following:

**A** Write your name.

**B** Read a page of this book.

**C** Count to 200.

**A1** How could you have made your measurements more accurate?

## Science in context: The Global Positioning System (GPS)

Global positioning systems work by receiving radio signals from satellites orbiting the Earth. Your GPS receiver calculates how far away each of these satellites is by using the time it takes radio signals to reach it. By knowing how far away you are from at least three fixed points, your position can be calculated accurately.

**7.2** *Network of GPS satellites.*

There are about 30 GPS satellites orbiting the Earth in known fixed positions. Wherever you are, you will be able to receive the signals from at least four of these. The accuracy of your position can be improved by using signals from other systems, such as the Russian GLONASS system (24 satellites) and the European system Galileo (24 satellites).

## Key facts:

- ✔ Rulers and measuring tapes can be used to measure distance.
- ✔ Clocks, watches, stopwatches and light gates connected to electronic timers, data loggers or computers can be used to measure time.
- ✔ Different pieces of apparatus give measurements with different levels of precision.

## Check your skills progress:

- I can choose suitable apparatus to measure distance and time.
- I can explain my choice of measuring apparatus.
- I can measure accurately and precisely and check that my results are reliable.

Chapter 7 . Topic 2

# Speed and average speed

You will learn:

- To calculate speed
- To represent scientific ideas using symbols and formulae
- To make predictions based on scientific knowledge and understanding
- To plan investigations, including fair tests, while considering variables
- To choose experimental equipment and use it correctly
- To evaluate the reliability of measurements and observations
- To take accurate measurements and explain why this matters
- To use results to describe the accuracy of predictions
- To carry out practical work safely
- To collect and record observations and measurements appropriately
- To describe results in terms of any trends and patterns and identify any abnormal results
- To present and interpret scientific enquiries correctly
- To describe the application of science in society, industry and research
- To discuss the global environmental impact of science

## Starting point

| You should know that... | You should be able to... |
|---|---|
| Some apparatus is more accurate and precise than others | Select appropriate equipment to measure distance and time |
| Gravity is a force that attracts objects towards the Earth | Measure distance and time accurately |
| Friction, including air resistance, is a force that slows down moving objects | Calculate the average (mean) of a set of numbers |

## Measuring speed

When you talk about an object's **speed**, you are describing how far it travels in a given time. Speed is measured in metres per second (m/s). In equations, speed is shown by $v$. A speed of 10 m/s means the object travels 10 m in 1 s.

### Key term

**speed**: how far something moves in a given time.

How far does a car with a speed of 12 m/s travel in:

**a)** 10 s?

**b)** 1 minute?

**c)** 10 minutes?

### Key term

**average**: the mean average of a set of numbers is found by:

$$\frac{\text{total of all the numbers added together}}{\text{how many numbers there are}}$$

As you travel to school your speed changes many times. We often measure the **average** speed of the moving object during a journey. This takes account of the object speeding up, slowing down and even stopping during its journey.

$$\text{average speed} = \frac{\text{total distance travelled}}{\text{total time taken}}$$

This can be written as $v = \frac{s}{t}$

where

$v$ = speed

$s$ = distance

$t$ = time

A girl takes 25 s to run 100 m. What is her average speed?

A man walks 450 m at an average speed of 3 m/s. How long is he walking for?

## Activity 7.3: Measuring your average walking speed

**A** Measure out a distance of between 5 and 10 metres.

**B** Use a stopwatch to time how long it takes you to walk this distance.

**C** Calculate your average speed.

**A1** You timed how long it takes to walk a set distance. Describe a different way of doing this experiment.

**A2** Why is it better to do this experiment with longer distances and times than shorter distances and times?

In science we often measure speed in metres per second, but in everyday life speed is often measured in different units, for example kilometres per hour (km/h) (see figure 7.3).

**4** Estimate the average speed of the following:

(You may find it useful to start by estimating the distance travelled and the time it would take to travel that distance.)

**a)** an Olympic sprinter running 100 m in m/s

**b)** a tortoise in cm/s, then convert your answer to m/s

**c)** a snail in mm/s, then convert your answer to m/s

**d)** a cheetah running at full speed in km/h, then convert your answer to m/s.

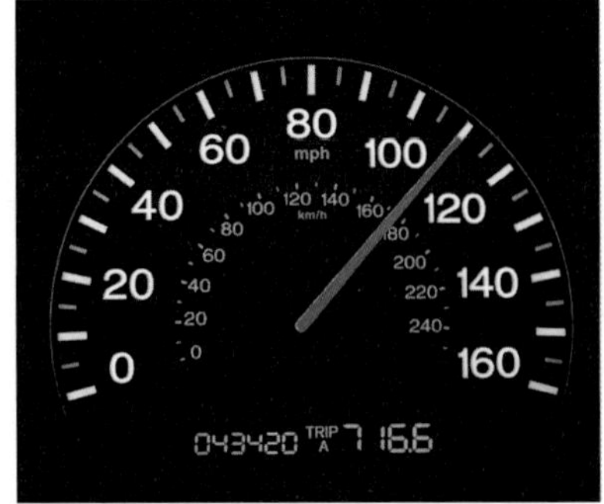


**7.3** *Speedometer.*

## Using light gates

Light gates are electronic sensors. They can measure time very accurately. In figure 7.4 the timer starts when the front of the card passes through the light gate and stops when the end of the card passes through it. This tells you how long it took the card to travel through the light gate. If the card is 10 cm long, you now know how long it took the car to travel 10 cm.

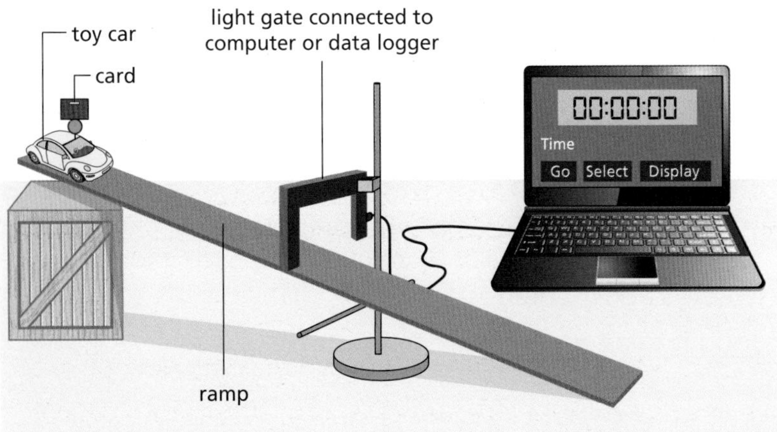


**7.4** *Using a light gate to measure time.*

**5** How could you use two light gates to measure how long it takes a toy car to travel 50 cm?

**Anomalous results** are results that do not follow the pattern of other results. You are less likely to get anomalous results if you repeat your experiment more than once. Anomalous results can affect your average values if they are very much higher or lower than the rest of your results.

### Key term

**anomalous results**: results which don't fit the pattern of the other results obtained.

**6** Ali timed how long it took a toy car to roll down a ramp. Here are his results.

10 s, 12 s, 14 s, 24 s, 11 s, 13 s, 10 s, 11 s, 12 s, 10 s

**a)** Which result is anomalous?

**b)** What is the average time if the anomalous result is included?

**c)** What is the average time if the anomalous result is not included?

## Making links

You may have studied both energy and friction earlier in your course and know that energy is dissipated (lost) when it is transferred from one object to another.

You can apply this knowledge to moving objects.

Jo pushes a toy car across a flat surface and measures its speed. She notices that the car slows down and finally stops.

Explain why.

## Activity 7.4: Measure the average speed of a toy car descending a ramp (1)

Use one of these methods to find the average speed of a toy car as it rolls down a ramp.

**Method 1: With a stopwatch**

**A** Time how long it takes a toy car to go from the top to the bottom. Repeat at least three times, discard any anomalous results, then calculate the average time.

**B** Now calculate the average speed of the car.

**Method 2: With a light gate (see figure 7.4)**

**A** Attach a card to the top of a toy car. Set up a light gate halfway along the ramp.

**B** Release the car at the top of the ramp and record the time taken for the card to travel through the light gate. Repeat at least three times, discard any anomalous results, then calculate the average time.

**C** Now calculate the average speed of the car.

**A1** Why is it important to place the light gate *halfway* along the ramp in method 2?

## Activity 7.5: Measure the average speed of a toy car descending a ramp (2)

**A** Plan an experiment to investigate how the average speed of a toy car depends on the steepness of the ramp.

**B** Design a table to record your results.

**C** Consider how to display your results as a graph.

**A1** What do you expect your experiment to show?

**D** Check your plan with your teacher before doing your experiment.

**A2** Did you get any anomalous results? If so, can you think of a reason for this?

### Making links

You may have studied gravity earlier in your course and know that the Earth's gravity attracts objects towards the Earth.

Use your knowledge of gravity to explain your results to the experiment in Activity 7.5.

## Science in context: Tracking migrating animals

GPS tracking can be used to track the movement of birds and land animals. A tiny GPS receiver is placed on the animal, and this receiver picks up signals from GPS satellites. The GPS data collected is sent to scientists via another satellite. The GPS receivers are small enough to attach to birds. For example, birdwatchers can see the tracks of ospreys as they migrate from Loch Garten in Scotland to the west coast of Africa. Scientists can use the data to try to discover why the animals are moving and also to discover the reasons for any changes in migration patterns.

**7.5** *The migration routes of ospreys.*

## Key facts:

- ✔ Average speed = distance travelled / time taken.
- ✔ Speed is measured in metres per second (m/s).

## Check your skills progress:

- I can plan an experiment to measure average speed.
- I can calculate the average speed of a moving object.
- I can make sure my results are reliable.
- I can identify anomalous results.

# Distance/time graphs

You will learn:

- To interpret and draw simple distance/time graphs
- To carry out practical work safely
- To collect and record observations/ measurements appropriately
- To take accurate measurements and explain why this matters
- To describe trends and patterns in results

## Starting point

| You should know that... | You should be able to... |
|---|---|
| Speed is measured in metres per second (m/s) | Select appropriate equipment to measure distance and time |
| Average speed = $\frac{\text{distance travelled}}{\text{time taken}}$ <br> $v = \frac{s}{t}$ | Draw line graphs with sensible axes, labels and scales |
| | Find the gradient of a graph |

## What are distance/time graphs?

A distance/time graph shows if an object is moving and if its speed is constant, increasing or decreasing.

The distance/time graph in figure 7.6 shows how long it takes a runner to run 10 m, 20 m and 30 m.

1 How can you tell that the runner is moving at a constant speed?

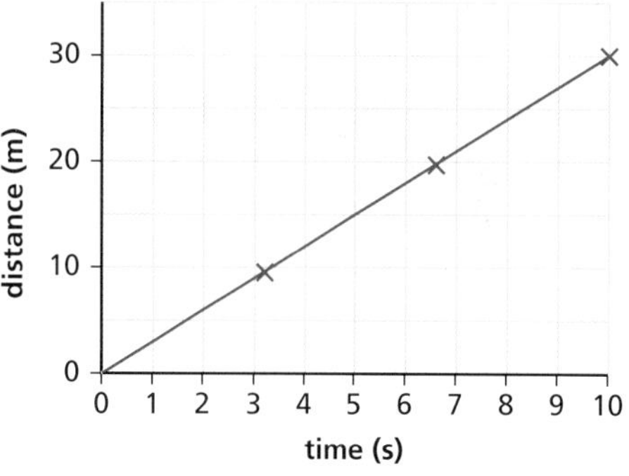


**7.6** *Distance/time graph showing a constant speed.*

## Calculating speed

To work out the average speed of the runner in figure 7.6 after 10 s:

$$\text{average speed} = \frac{\text{distance travelled}}{\text{time taken}} = \frac{30\text{ m}}{10\text{ s}}$$

$$= \frac{s}{t}$$

This is the same as working out the **gradient** (slope) of the graph.

### Key term

**gradient**: the gradient of a graph tells you how steep the line is.

$$\text{gradient} = \frac{\text{distance travelled}}{\text{time taken}}$$

For the runner in figure 7.6:

$$\text{gradient} = \frac{30\text{ m}}{10\text{ s}}$$

Figure 7.7 shows a distance/time graph for a faster runner.

**2** What is the speed of the runner in figure 7.7?

**3** Graphs 7.6 and 7.7 both have the same axes. Which graph has the steeper slope – the one for the faster runner or the one for the slower runner?

*7.7 Distance/time graph for a faster runner.*

A distance/time graph can tell you a lot about what an object is doing.

Figure 7.8 shows part of a cyclist's journey. In part A, the cyclist moves at a constant speed. In part B he is still moving at a constant speed, but now he is moving faster. In part C he slows down.

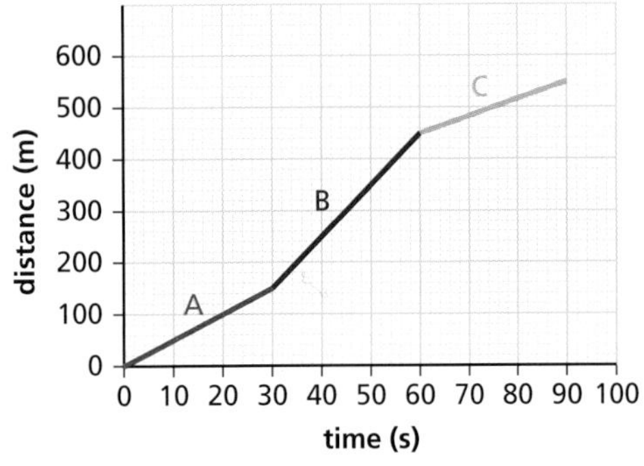


**7.8** *Distance/time graph for moving at different average speeds.*

**4** Look at figure 7.8.

**a)** Explain how you know that the cyclist is travelling faster in part B than in part A.

**b)** Is the cyclist's speed in part C faster or slower than part A? How do you know?

Figure 7.9 is a distance/time graph of an animal that runs at a constant speed for 3 s and then stops.

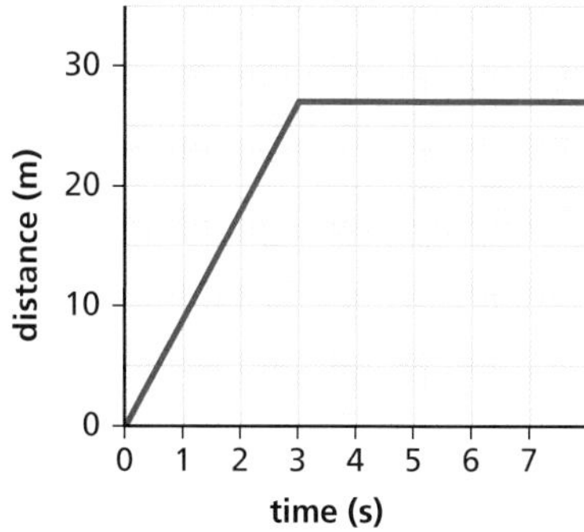


**7.9** *Distance/time graph of an animal that runs at a constant speed for 3 s and then stops.*

**5** How do you know that the animal in figure 7.9 stopped running after 3s?

The graph in figure 7.10 shows a distance/time graph for an animal whose speed is changing all the time.

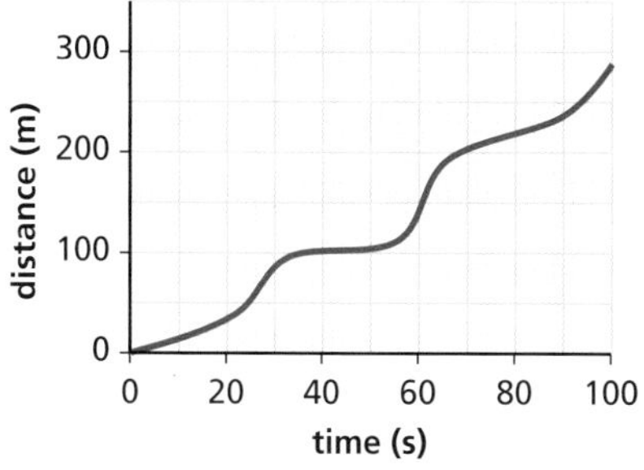


**7.10** *Distance/time graph of an animal whose speed is changing all the time.*

**6** How can you tell from the graph in figure 7.10 that the animal's speed is changing all the time?

**7** Figure 7.11 shows the distance/time graph of a walk. Match the labelled sections (A–E) to the descriptions here.

I am walking back towards the start of my walk.
I am walking at a constant speed.
I am tired so I am walking more slowly.
I am walking a little quicker.
I am walking downhill so my speed increases.

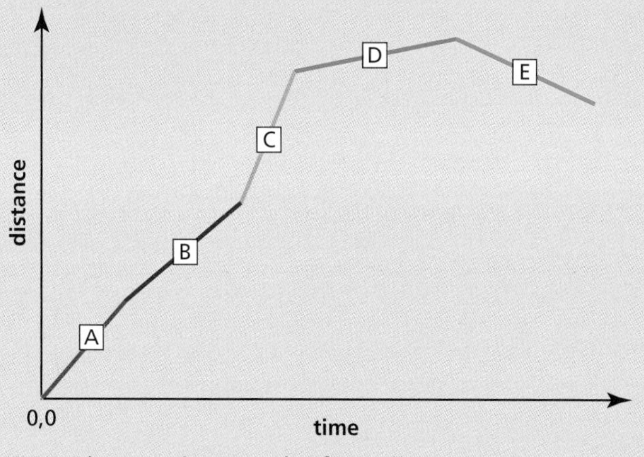


**7.11** *Distance/time graph of a walk.*

## Plot your own distance/time graph

### Activity 7.6: Go on a journey

Work in a small group.

**A** Place four markers on the ground, 10 m apart (see figure 7.12).

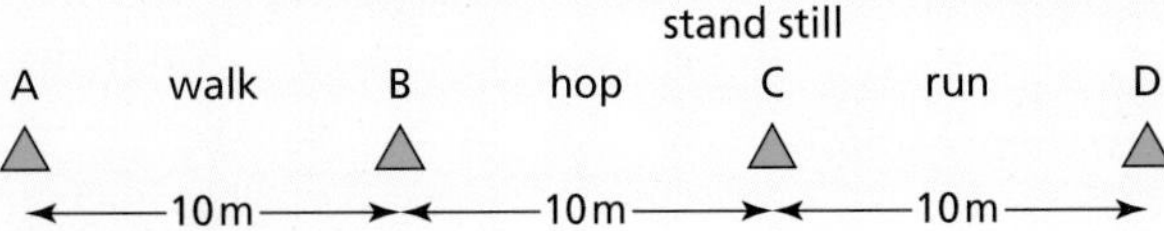


**7.12** *Timing a journey.*

**B** One of you:

- walks from A to B
- then hops from B to C
- then stands still for 5 s
- then runs from C to D.

Time how long each stage of the journey takes. Record the results in a table.

**A1** Work out the average speed for each stage of the journey. Draw a distance/time graph of the journey.

### Key facts:

✔ The gradient of a distance/time graph tells you the speed of a moving object.

✔ The steeper the gradient of a distance/time graph, the faster an object is moving.

✔ A straight horizontal line on a distance/time graph shows that the object has stopped moving.

### Check your skills progress:

- I can draw distance/time graphs for moving objects.
- I can describe the motion shown in a distance/time graph.

# End of chapter review

## Quick questions

1. How far will a train with a speed of 20 m/s travel in:

   (a) 10 s?

   (b) 2 minutes?

   (c) 10 minutes?

2. A dog takes 20 s to run 220 m. What is its average speed?

3. How far could a horse with an average speed of 15 m/s run in:

   (a) 10 s?

   (b) 1 minute?

4. How long would it take a leopard, running at an average speed of 20 m/s, to travel 500 m?

5. Speeds can be measured in different units.

   (a) What is a speed of 3 m/s equal to in cm/s?

   (b) What is a speed of 50 cm/s equal to in m/s?

   (c) What is a speed of 60 km/h equal to in m/s? Give your answer to the nearest whole number.

6. What is the average speed of the object shown in the graph below?

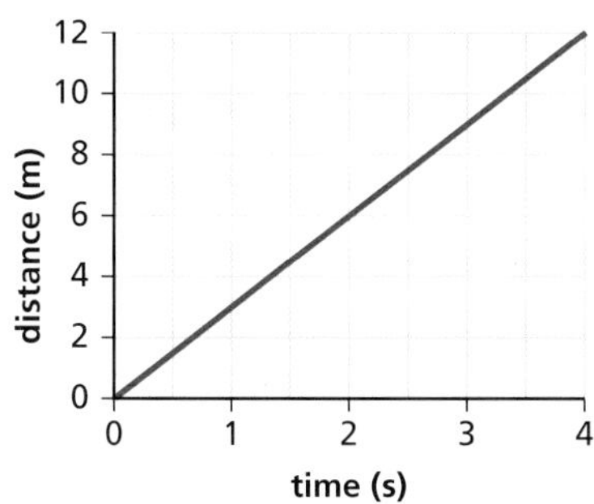


## Connect your understanding

7. Copy the graph in Question 6. Now add a line showing an object moving at an average speed of 2 m/s.

8. Why is a light gate a more accurate way of measuring time than a stopwatch?

9. The graph below shows the average speed of a racing car at different times as it travels around part of a race circuit. There is a 10 s time interval between each reading. Average readings over each 10 s interval have been plotted.

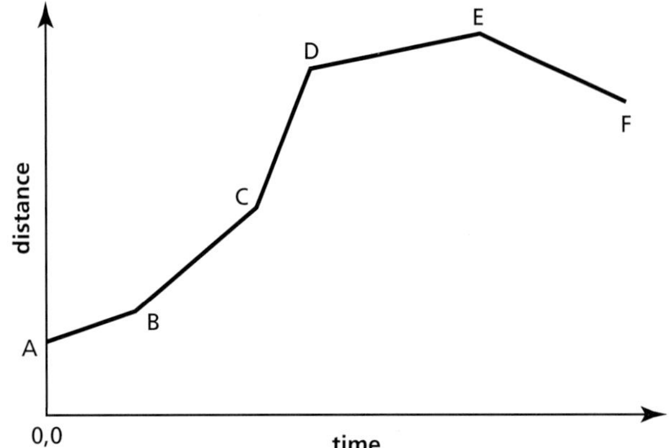


**(a)** Does the graph show the start of the race? How do you know?

**(b)** The car slows down when it reaches a corner. Which point on the graph shows where a corner could be?

**(c)** Which point on the graph shows the car at the furthest point from the start of the race? How do you know?

**(d)** Challenge What would the graph look like if the average speed were measured every 1/10th second rather than every 10 seconds? Give a reason for your answer.

10. Sketch distance/time graphs for the following:

**(a)** A car travelling away from its starting point at a constant speed, then braking and coming to a stop.

**(b)** A bike moving at a constant speed, then suddenly stopping.

11. A student wants to find the average speed of a beetle. She measures how far the beetle moves every 10 s for 1 minute. Her results are shown in the table.

| Time (s) | Distance (m) |
|---|---|
| 0 | 0 |
| 10 | 4 |
| 20 | 8 |
| 30 | 12 |
| 40 | 15 |
| 50 | 20 |
| 60 | 24 |

(a) Draw a distance/time graph of these results.

(b) What is the beetle's average speed over one minute?

(c) Which one of her readings suggests that the beetle is not moving at a constant speed? How can you tell this from the graph?

## Challenge question

12. A student is measuring the time it takes a ball to fall from a height of 2 m to the ground. Her results are shown in the table.

| Time (s) |
|---|
| 0.65 |
| 0.67 |
| 0.65 |
| 0.56 |
| 0.66 |
| 0.64 |
| 0.65 |

(a) Which of these results is anomalous?

(b) Calculate the mean time it takes for the ball to fall to the ground. Use all seven results.

(c) Calculate the mean time it takes for the ball to fall to the ground. Use only the results that are not anomalous.

(d) Which of the calculations, (b) or (c), is likely to give the more accurate answer?

(e) Calculate the average speed of the ball as it falls to the ground. Use the more accurate value of mean time taken to reach the ground.

8

# Chapter 8

Forces

## What's it all about?

A mountain biker depends upon gravity to accelerate to high speeds as they go down steep slopes. Their speed also depends upon other forces: the force they supply to the wheels from the pedals and friction between the tyres and the ground. Snow or ice change dramatically the friction forces and how the bicycle responds. In snow, wider tyres are needed so that the biker is less likely to sink into the snow. To work out where the bicycle will move to, the biker instinctively adds up the continually changing forces and steers and accelerates or brakes. The difference between setting a record time and having a massive crash rests on split-second decisions.

You will learn about:

- Balanced and unbalanced forces
- How turning forces are caused
- How pressure in gases and liquids is caused by particle movement, and how to calculate pressure
- How diffusion is explained by the particle model

You will build your skills in:

- Describing and using analogies
- Planning a method for an investigation while considering variables to change, measure or control

Chapter 8 . Topic 1

# Balanced and unbalanced forces

You will learn:

- To describe how forces affect motion
- To represent scientific ideas using symbols and formulae
- To describe how evidence affects scientific hypotheses
- To make predictions based on scientific knowledge and understanding
- To plan investigations, including fair tests, while considering variables
- To collect and record observations and measurements appropriately
- To evaluate the reliability of measurements and observations
- To use results to describe the accuracy of predictions
- To reach conclusions studying results and explain their limitations
- To describe the application of science in society, industry and research
- To discuss the global environmental impact of science

## Starting point

| You should know that... | You should be able to... |
|---|---|
| An object may have forces acting upon it, even when at rest | Use force diagrams to show the name, direction and size of forces acting on an object |
| With no force, a stationary object stays still | Plan an investigation to test an idea |
| If a force is applied to an object it changes speed or direction | Identify control variables in investigations |
| Friction and air resistance reduce the speed of moving objects | Explain why it is good to repeat measurements |

## Effects of forces

You know that forces can change the way objects are moving by making them speed up, slow down or change direction.

- The force of gravity makes a stone speed up when you drop it.
- Friction between two solid surfaces slows an object down or stops it moving, such as when you go down a slide with a rough surface. Friction with the air slows down objects moving through air. This is called air resistance. Parachutists use air resistance to slow them down to land safely. Opening the parachute increases air resistance and slows the fall.
- A strong wind can blow a plane off course.

1. Describe how the effects of friction and air resistance can be reduced.
2. Would lubricating the road surface to reduce friction make a car travel faster?

## Adding forces

Sometimes more than one force acts on an object. You can add these forces together to make a single force, called the **resultant force**. This resultant force shows you the combined effect of all the different forces on an object.

You have seen how to draw forces on diagrams. Drawing a diagram helps show how forces add up.

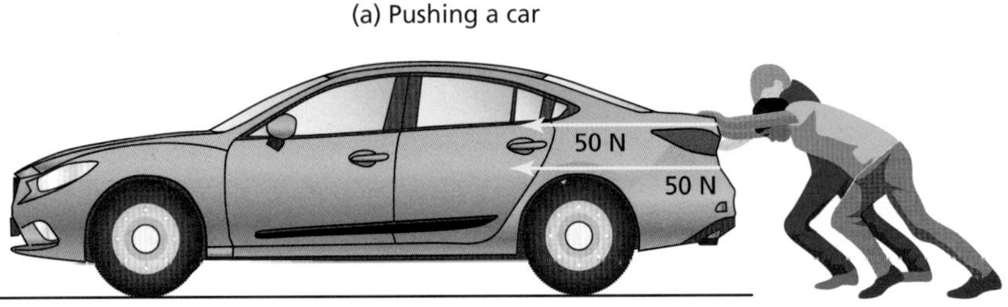


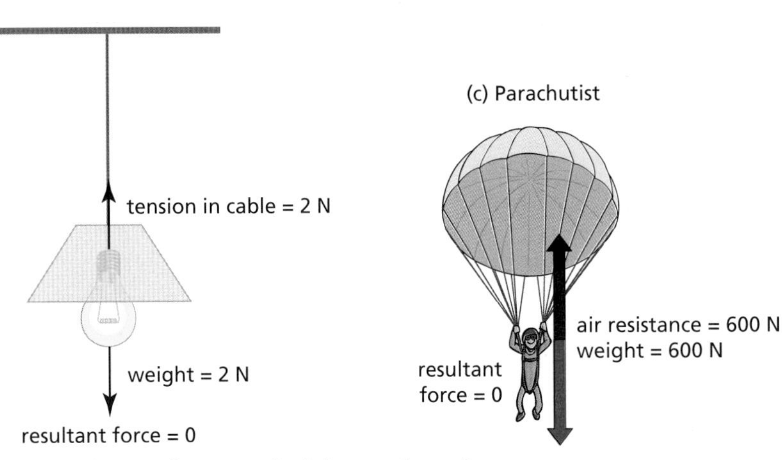


**8.1** *Combining forces to find the resultant force.*

When there is more force in one direction than the other, the forces are **unbalanced** and there is a resultant force. A resultant force will change how an object is moving. It could make it start to move, stop moving or change speed or direction.

Sometimes the total force in one direction is equal to the total force in the opposite direction and the resultant force is zero. When this happens, we say that all the forces on an object are **balanced forces**.

### Key terms

**balanced forces**: when the resultant force is zero.

**resultant force**: shows the single total force acting on an object when all the forces acting on it are added up.

**unbalanced forces**: when there is a resultant force.

So, that object either stays at rest, or if it is moving already it keeps a constant speed and direction. For two forces to balance they need to be equal in size but opposite in direction.

**3**

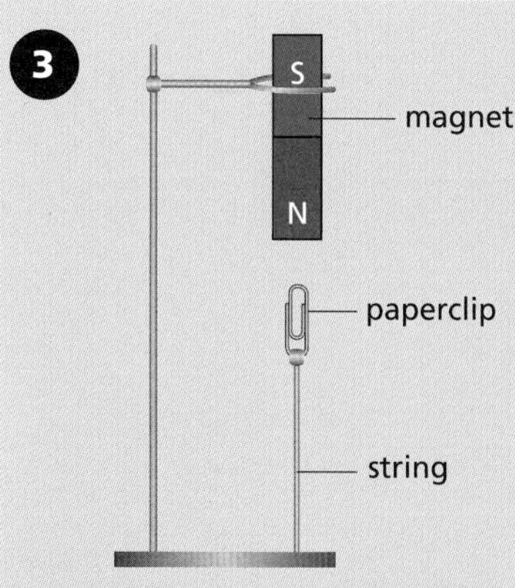


8.2

**a)** Name the two forces acting on the paperclip.

**b)** Copy these sentences and choose the correct ending: The paperclip doesn't move. This means (the downward force is biggest/ the upward force is biggest/ the upward and downward forces are equal).

**c)** Draw the paperclip and add arrows to show the size and direction of the two forces.

**4** For the following situations:

(i) Draw a force diagram with arrows showing the size and direction of each of the forces.

(ii) Find the resultant force.

(iii) Say what will happen.

**a)** Hanna is cycling. Her legs provide a forward force of 150 N. Friction and air resistance give a combined force of 150 N.

**b)**

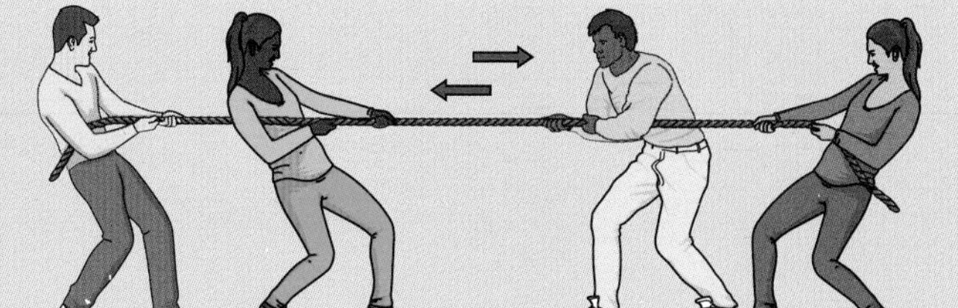


8.3

Team A pull with a total force of 500 N and team B pull with a total force of 450 N.

**c)** A car's engine provides a forward force on the car of 700 N. Friction and air resistance produce a combined backward force of 400 N.

**5** Calculate the resultant force on this box.
Hint: find the resultant of the two forces moving to the right first.

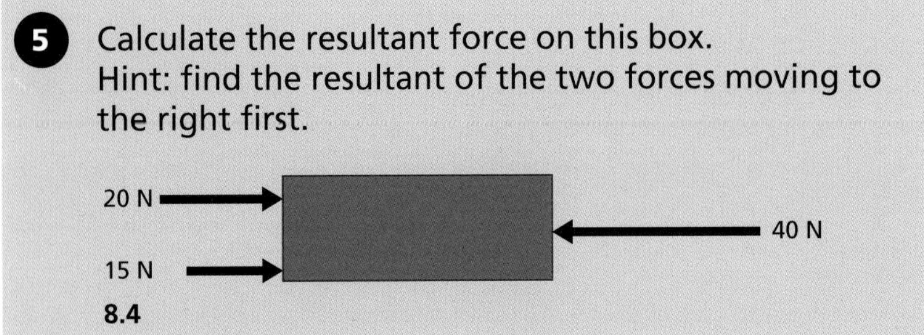


**8.4**

## Science in context: Reducing energy and fuel needed by cars

The air resistance on a moving car increases as the speed increases. The greater the air resistance, the more fuel the car needs to burn to overcome this force and keep moving at a steady speed. Making cars streamlined helps to reduce the use of environmentally damaging fossil fuels. Driving at a lower speed also reduces the amount of fuel used and the pollutants emitted.

**8.5** *A streamlined car moves through the air with less resistance, so wastes less energy and is more fuel-efficient.*

**6** Draw and label force diagrams to show the forces on a streamlined car and an unstreamlined car moving at the same high speed. Use the diagrams to explain why the engine in the unstreamlined car has to exert more force to travel at the same constant speed.

## Floating and sinking

When an object is placed in water, the water pushes upwards on the object. The upwards force from the water is called **upthrust**. There is also an upthrust on objects in air.

### Key term

**upthrust**: the upwards force from a liquid on an object in a liquid. Also applies to the upwards force on an object in a gas.

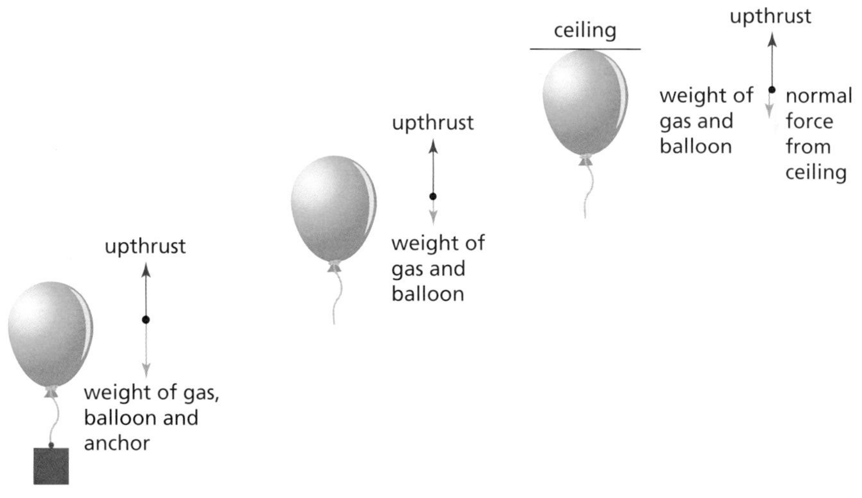


**8.6** *a) floating in mid air* *b) rising up* *c) stationary when it hits the ceiling*

Imagine a balloon inflated with helium. If the balloon is released it rises because there is an upthrust force from the air around it. The weight of the gas (and balloon) is not enough to balance the upthrust.

To stop helium balloons rising up, weights are often attached (figure 8.6a).

With the weight attached, the forces are balanced and the resultant force is zero. The balloon stays still. If the weight is removed (figure 8.6b), the resultant force is upwards so the balloon starts to move up. When it reaches the ceiling, there is a downward force on the balloon from the ceiling. This is called the **normal force**. The combination of the downwards forces is equal to the upthrust so the resultant is once again zero.

## Key term

**normal force**: when an object touches a surface there is a force on the object at 90° to the surface (normal to the surface).

7. A girl weighing 400 N floats in a swimming pool. What size is the upthrust force on her?

8. Juan is cycling at a steady speed, using a force of 200 N. He stops pedalling and soon the bicycle stops. Draw a force diagram for Juan and his bike when he stops pedalling. Use your diagram to explain why he stops.

9. Jenni swims under water holding an inflatable toy. She lets go of the toy at the bottom of the pool. The upthrust on the toy is greater than its weight. Draw a force diagram for the toy when it is released. Use your diagram to explain how the toy moves.

## Activity 8.1: Testing a hypothesis about friction

Khalil watched a dumper truck unloading. He noticed that the truck must be lifted higher to unload some materials than others.

Khalil thought this observation could be explained if there is less friction when the truck is tipped more.

**8.7** *A dumper truck unloading.*

**A1** Write Khalil's **hypothesis** in a way which could be tested in an experiment. Use your scientific knowledge to give an explanation for your hypothesis.

**A2** Design an investigation to test this hypothesis.

When planning your investigation you will need to consider:

- What you will change (independent variable).
- What you will measure (independent variable).
- What you will keep the same (control variables).
- How you will check that your results are **reliable**.

**A3** Draw a table for your results. Allow for repeat measurements and calculating the mean of your repeats.

**A4** When your teacher has checked your method, carry out the experiment and record your results.

**A5** Write a conclusion. Do your results support the hypothesis?

**A6** Write an evaluation of your experiment. Are your results reliable? How do you know?

### Key terms

**hypothesis** (plural hypotheses): a statement or claim that can be tested using experiments. It is proved or disproved by scientific enquiry.

**reliable**: measurements are reliable when repeated measurements give results that are very similar.

### Key facts:

- ✔ The forces on an object combine to give an overall resultant force.
- ✔ A resultant force of zero means all the forces on an object are balanced.
- ✔ Balanced forces on an object result in steady motion in the same direction, or staying at rest.
- ✔ Unbalanced forces on an object change its speed or direction of motion.

### Check your skills progress:

- I can make and explain predictions based on scientific knowledge.
- I can plan an investigation to test a hypothesis.
- I can identify control variables in an investigation.
- I can explain how I would get reliable results.

Chapter 8 . Topic 2

# Turning effect of a force

You will learn:

- To identify and calculate turning forces
- To represent scientific ideas using symbols and formulae
- To make predictions based on scientific knowledge and understanding
- To collect and record observations and measurements appropriately
- To describe results in terms of any trends and patterns and identify any abnormal results
- To reach conclusions studying results and explain their limitations
- To use results to describe the accuracy of predictions
- To evaluate experiments and investigations, and explain any improvements suggested

## Starting point

| You should know that... | You should be able to... |
|---|---|
| Force diagrams are used to show the name and direction of forces acting on an object | Describe how scientific hypotheses can be supported or contradicted by evidence from an enquiry |
| On a force diagram the direction and size of the force is shown by an arrow | |
| An object may have many forces acting on it, even when at rest | |

## Turning forces

Turning forces (also known as **moments**) make an object turn either clockwise or anticlockwise. To have a turning force you need something to apply the force (**effort**) and something for the force to turn around (**pivot**).

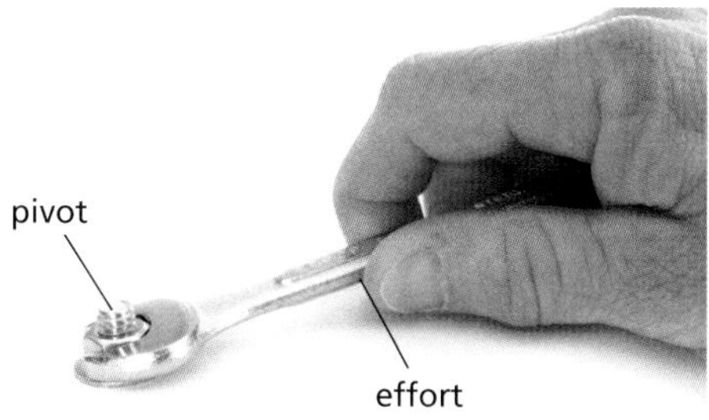


**8.8** *A spanner can be used to turn a nut. The effort comes from the hand. The pivot is at the nut.*

### Key terms

**effort**: the force that you put in to move the load.

**moment**: turning effect of a force – measured in newton-metres.

**pivot**: the point about which the force turns.

### Uses for turning forces

It is difficult to turn a nut using your hand. A spanner makes it easier to turn a nut because the effort is applied at a distance from the pivot. This produces a larger turning effect – a larger moment.

The idea of moments is also useful when a lever is used to lift an object by turning it about a pivot. Levers help us to move large **loads** using a smaller effort – door handles make it easier to turn the mechanism to open the door.

8.9 *Door handles make use of turning forces.*

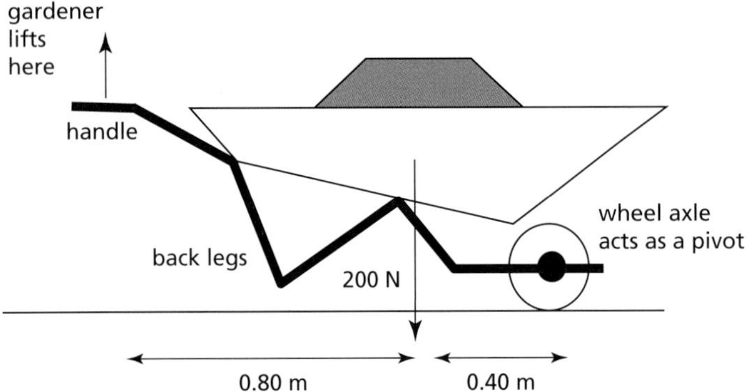


**8.10** *A wheelbarrow can be used to lift heavy loads*

A wheelbarrow also makes it easier to lift heavy loads. With a wheelbarrow, the load is on the same side of the pivot as the effort.

### Key term

**load**: the force that you need to move.

### Making links

Topic 1.3 describes how muscles move joints in our bodies. The joints in our bones are pivots. We have muscle pairs around each joint in our body.

Explain how the pairs of muscles in our arm lift up the weight in figure 8.11.

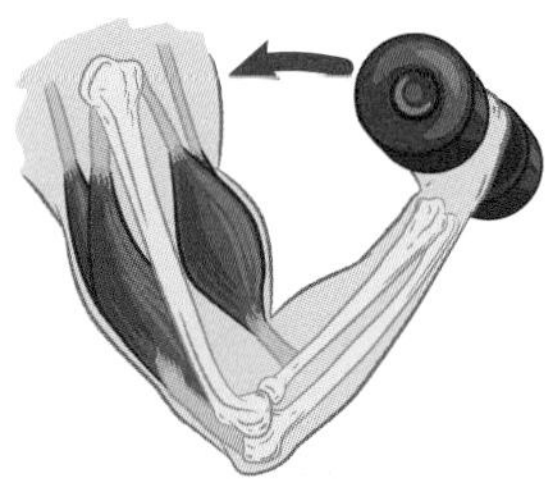

**8.11** *Using arm muscles to lift a weight.*

## Activity 8.2: Finding turning forces at home

Many objects in your home use turning forces – find some objects like this in your home.

**8.12** *Using a spanner.*

**A1** Draw the objects you have found – label the pivot and where you apply the effort.

**A2** Apply a force to the object. Describe what happens.

**A3** Apply a force to the object, but closer to the pivot. Compare how difficult it is to use turning forces closer to the pivot with further away from the pivot.

**A4** Do your observations support or contradict the hypothesis that increasing the distance of the force from the pivot increases the turning effect of the force?

**A5** How could you test this hypothesis? Think about the independent variable, the dependent variable and the control variables.

## Calculating turning forces

A turning force can be calculated by multiplying the force by its distance from the pivot:

moment = force × distance from the pivot

Using symbols, this can be written as $M = F \times s$. You may remember from Topic 7.2 that the symbol $s$ is for distance. We use the symbol $F$ for a force.

Moments are measured in newton-metres (N m). This is because when you multiply force (measured in newtons) and distance (measured in metres), you also multiply their units. Multiplying newtons and metres gives newton-metres.

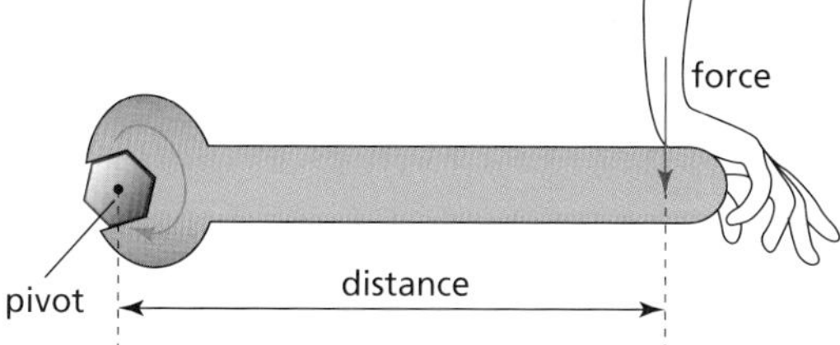


**8.13** *You calculate the moment of a force by multiplying the force and the distance.*

1. Use the terms *pivot*, *effort* and *load* to explain why it is easier to turn a nut using a spanner rather than your hand. Compare the similarities and differences between the pivot, effort and load in each case.

2. Nori uses a spanner to turn a nut. He uses a force of 20 N to turn a wrench that is 0.2 m long. What moment does he apply?

3. Toyun lifts a wheelbarrow full of sand by pulling upwards on the handles with a force of 200 N. The handles are 0.5 m from the wheel (the pivot). What moment does Toyun apply about the pivot?

## Balancing moments

If two moments are acting on opposite sides of a pivot, we can calculate whether they are balanced or not. One moment is described as 'anticlockwise' because it turns around the pivot in the anticlockwise direction. The other moment is the clockwise moment because it turns around the pivot in the clockwise direction.

First, we need to calculate the moment of each force. If the moments are equal then the moments are balanced. This is called the principle of moments:

When an object is balanced, the sum of the anticlockwise moments is equal to the sum of the clockwise moments.

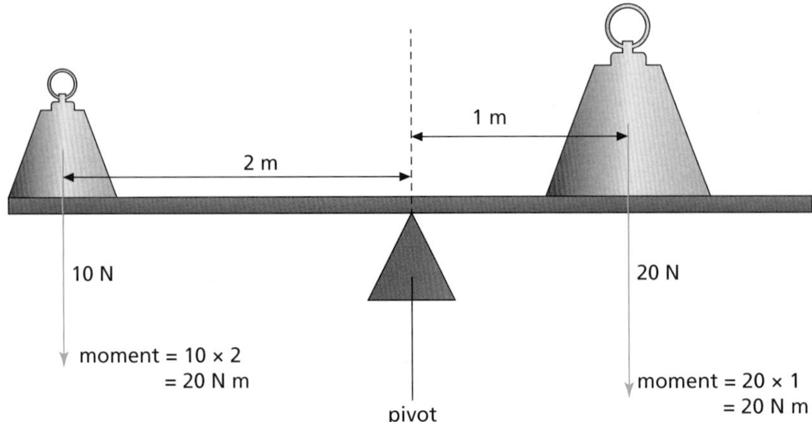


**8.14** *These moments are balanced.*

For example, Chen and Nuo sit on a see-saw. Chen weighs 600 N and sits 1 m away from the pivot. Nuo weighs 400 N and sits 1.5 m away from the pivot. Are they balanced?

moment = force × distance from the pivot

Chen's moment = 600 × 1 = 600 N m

Nuo's moment = 400 × 1.5 = 600 N m

So Chen and Nuo are balanced.

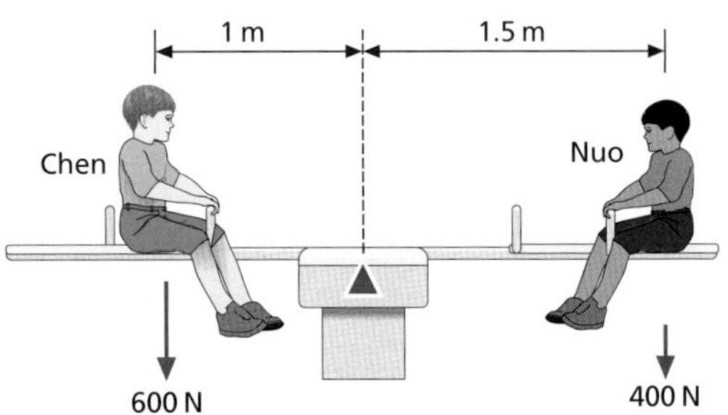


**8.15** *Is the see-saw balanced?*

Yejida and Zane sit on opposite sides of a see-saw. Zane weighs more than Yejida. Zane realises that since he weighs more, to balance Yejida he must sit closer to the pivot.

This is because moment = weight × distance from the pivot. So if you want to keep the moment the same, for a bigger force you have to decrease the distance from the pivot.

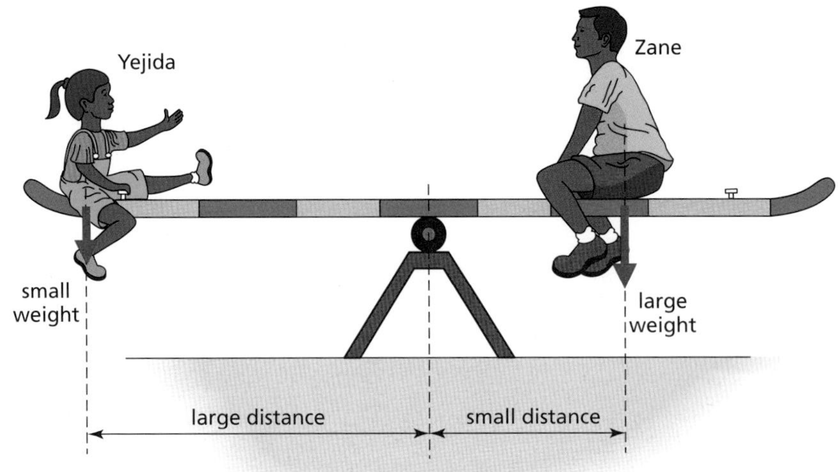


**8.16** *Zane has to adjust his position on the see-saw as he has a different weight from Yejida.*

When two or more forces are on the same side of the pivot, the moment on that side is equal to the sum of all the moments. For example, Anil and Mythri sit on the same side of the see-saw. Anil weighs 500 N and sits 2 m from the pivot. Mythri weighs 350 N and sits 1 m from the pivot.

Anil's moment = 500 × 2
= 1000 N m

Mythri's moment = 350 × 1
= 350 N m

**8.17** *There are two anticlockwise moments and one clockwise moment.*

Their total moment = 1000 + 350 = 1350 N m

Mark sits 2 m away from the pivot. As the see-saw is balanced, the anti-clockwise and clockwise moments on the see-saw must be balanced.

The total moment for Anil and Mythri must be equal to Mark's moment.

If Mark's moment is equal to 1350 N m, and he is sitting 2 m away from the pivot, he must have a weight of 675 N.

Force = moment/distance

Force = 1350/2

Force = 675 N

**4** Inga sits on a see-saw. Mazin sits on the other side of the pivot. Draw a labelled diagram showing the turning forces acting on the see-saw.

**5** Olamide has a weight of 300 N. Amara has a weight of 200 N. Olamide and Amara both sit on the same side of a see-saw. Olamide sits 1.2 m from the pivot and Amara sits 2.5 m from the pivot. Calculate the total moment of Olamide and Amara.

**6** Chandra has some balance scales in her kitchen. She puts an apple that weighs 1.3 N in one pan, 0.2 m from the pivot. She puts some masses with a weight of 0.8 N in the other pan, 0.3 m from the pivot. Calculate the moments and determine if the scales are balanced.

**8.18** *Balance scales.*

Cranes are designed to lift heavy weights and move them from one place to another. Cranes can be used to move large items to and from ships.

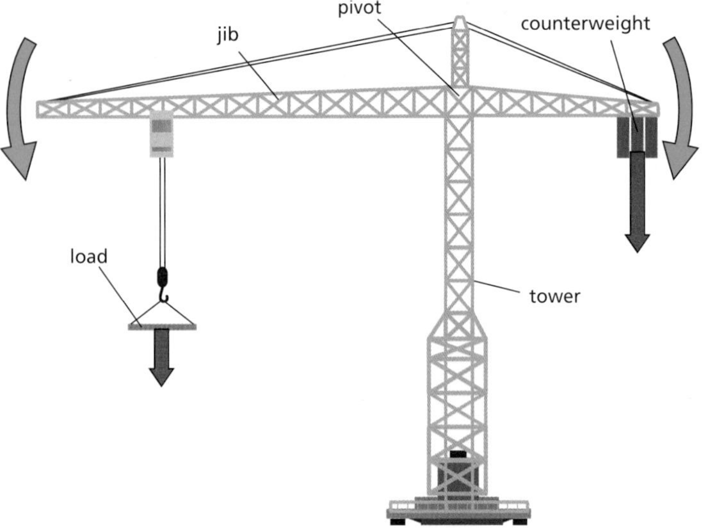


**8.19** *The pivot, load and counterweight on a crane.*

The load is usually at a large distance from the pivot along the metal arm called the jib. To stop the crane toppling over, a counterweight is placed on the other side of the pivot. To balance the moments the counterweight must be heavier than the load, because it is closer to the pivot.

## Key facts:

- ✔ You calculate moments by multiplying the force by the distance of that force from a pivot.
- ✔ Moments are measured in newton-metres (N m).
- ✔ Moments act in a clockwise or an anticlockwise direction.
- ✔ If the total clockwise moments and the total anticlockwise moments are equal then there is no turning effect.
- ✔ The principle of moments can be applied to devices such as levers and cranes.

## Check your skills progress:

- I can use the equation $M = F \times s$ to represent the idea that two things can balance when their moments are equal.
- I can plan an investigation to test a hypothesis.
- I can identify control variables in investigations.

Chapter 8 . Topic 3

# Pressure on an area

You will learn:

- To explain what causes pressure on an area
- To make predictions based on scientific knowledge and understanding
- To take accurate measurements and explain why this matters
- To carry out practical work safely
- To reach conclusions studying results and explain their limitations
- To evaluate experiments and investigations, and explain any improvements suggested
- To describe the application of science in society, industry and research

## Starting point

| You should know that... | You should be able to... |
|---|---|
| Weight is a force and is measured in newtons (N) | Measure mass in kg using scales |
| | Present results in a table, where each column has a heading that includes the units of the measurement |

## What is pressure?

**Pressure** is caused by a force pushing down on an area. The effect of the pressure depends on the size of the force and the size of the area over which the force is applied. An elephant exerts a large pressure because it has a large weight. A pin is also capable of exerting a large pressure because the force from the person pushing on it is spread over a tiny point area.

Pressure increases when the amount of force increases. For example, if you push the pin harder, the pressure will increase.

Pressure also increases when the area the force is spread over decreases. If an elephant stood on one leg, it would exert the same amount of force on the ground, but its pressure will be bigger because that force is spread over a smaller area.

### Key term

**pressure**: the amount of force per unit of area – usually measured in newtons per square metre.

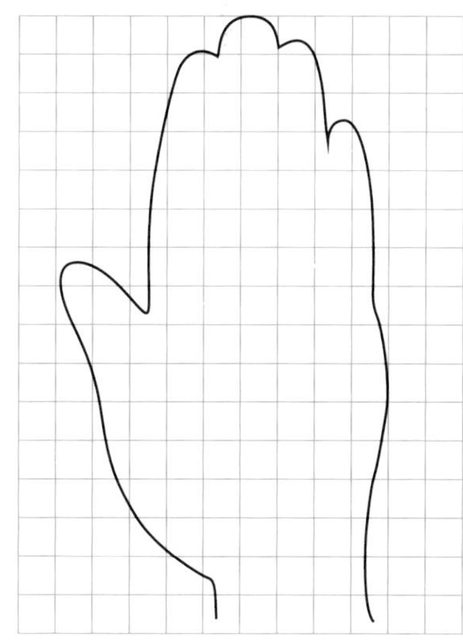

**8.20** *You can estimate the area of something by drawing around it on squared paper and counting how many whole squares it covers.*

## How do we calculate pressure?

To calculate the pressure, we divide the force by the area the force acts on:

$$\text{pressure} = \frac{\text{force}}{\text{area}}$$

Using symbols, this can be written as $p = \frac{F}{A}$

Since the units for force are newtons (N) and the units for area are metres squared ($m^2$), then the units for pressure are newtons per metre squared ($N/m^2$). If the area is in centimetres squared ($cm^2$), then the pressure is measured in newtons per centimetre squared ($N/cm^2$).

You calculate the area of an irregular shape by drawing around it on a piece of squared paper and counting how many centimetre squares it covers.

## Pressure calculations

A concrete block exerts a force of 15000 N over an area of 4 $m^2$. What pressure does it exert on the area?

$$\text{pressure} = \frac{15\,000}{4} = 3750 \text{ N/m}^2$$

The pressure is 3750 $N/m^2$. Another way of looking at this is shown in figure 8.21.

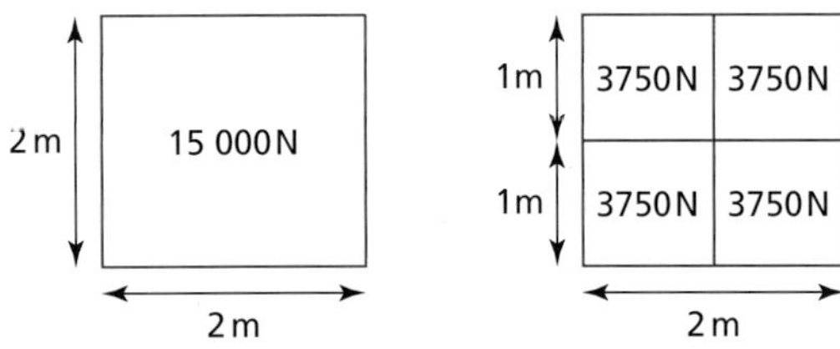


**8.21** *The concrete block exerts 15000 N of force over 4 squares, each of 1 square metre, to give a pressure of 3750 N/m².*

A pinhead has an area of 0.0000125 $m^2$. Someone pushes it into some wood with a force of 1 N. What pressure does the pinhead exert on the wood?

$$\text{pressure} = \frac{\text{force}}{\text{area}}$$

$$= \frac{1}{0.0000125} = 80000 \text{ N/m}^2$$

An elephant has a weight of 50000 N. It has four feet. Each foot has an area of 0.125 $m^2$. How much pressure does the elephant exert on the ground when it stands on all four feet? How much pressure does it exert on the ground when it stands on one foot?

Total area of elephant's feet = 4 × 0.125 = 0.5 $m^2$

$$\text{pressure} = \frac{\text{force}}{\text{area}} = \frac{50000}{0.5} = 100000 \text{ N/m}^2$$

Area of 1 foot = 0.125 $m^2$

$$\text{pressure} = \frac{\text{force}}{\text{area}} = \frac{50000}{0.125} = 400000 \text{ N/m}^2$$

1 Copy and complete this table.

| Force (N) | Area ($m^2$) | Pressure ($N/m^2$) |
|---|---|---|
| 10 | 2 | |
| 25 | 5 | |
| 15 | 3 | |
| 30 | 12 | |
| 40 | 10 | |
| 30 | 5 | |

**Table 8.1** *Calculating pressure.*

2 Isidor decides to measure the pressure an object exerts on sand by measuring the depth of the sand. What must Isidor keep the same?

3 Abel cuts a slice of bread with a knife. He decides that, because it was so easily cut, he must have used a lot of force. Is this correct?

4 Agnar wants to measure the pressure that two of his classmates exert on the floor when they stand on the floor. He measures Rupinder's foot area and her weight when she is wearing socks but no shoes; and he measures Kauri's foot area and his weight when he is wearing shoes. Why is this not a fair test?

## Activity 8.3: How much pressure do I exert?

You are going to calculate how much pressure your feet exert on the ground.

**8.22** *Bathroom scales measure your mass.*

**A** Weigh yourself standing on both feet.

**B** Scales give a reading in kilograms, which is a measurement of mass. You need your weight which is measured in newtons. To convert to newtons, multiply the mass in kilograms by 10 N/kg. (Remember, weight = mass × gravitational field strength.) Draw a suitable table and record your results.

**C** Now weigh yourself standing on one foot. Add the results to your table.

**D** Estimate the area of your shoes in centimetres squared ($cm^2$) by drawing around your shoes on some squared paper and counting how many whole centimetre squares are covered.

**A1** Is your weight any different to when you stand on both feet?

**A2** Calculate the pressure, in $N/cm^2$, exerted by your feet when you stand on both feet by dividing your weight by the area of your shoes.

**A3** Calculate the pressure exerted by your feet when you stand on one foot. How does this compare to the pressure you exert when standing on two feet?

**A4** Suggest how to improve the estimate of the area of your shoes.

## Why is pressure important?

### Knives, nails and pins

Knives cut better when they are sharp. A very sharp knife has the thinnest possible piece of metal in contact with the object it is cutting. The force is spread over the smallest possible area and so the pressure is higher.

The same principle applies to nails and pins. It is easier to push these into very hard walls because they have a tiny point. This means that the area in contact with the wall is very small and so the pressure of the pins or nails on the wall is very large.

### Feet

Animals such as camels that live in areas with soft ground have large feet. This spreads their weight over a large area and so the camels exert less pressure on the surface of the ground. This makes them less likely to sink and so it is easier for them to walk over the surface.

**8.23** *A camel's feet are big to reduce its foot pressure so it will not sink into the sand.*

### Tyres

We fit different tyres to vehicles depending on what type of surface they are moving on. If a vehicle is going over sand or mud, it may sink into the soft surface. The surface becomes more difficult to drive across. This is why vehicles that move over soft ground have wider tyres, and sometimes more wheels. The tyres spread the weight of the vehicle over a larger area. This reduces the pressure on the surface caused by the vehicle. The vehicle is less likely to sink.

**8.24** *A road bike (on the left) has narrower tyres than an off-road bike (on the right). This is because off-road bikes may go through mud and sand where they could sink. The wider tyres reduce their pressure so that they are less likely to sink.*

### Science in context: Artificial skin

Human skin is very sensitive and can detect very small changes in pressure. Scientists are developing artificial electronic skin that can sense touch and can be used on robots or for people whose skin is damaged. Electronic circuits detect how hard an object presses onto an area of the 'skin'. The challenge is to make this circuit thin and flexible. The research teams need to work with circuit designers and engineers who design new materials.

5 What size tyres would you put on a bicycle to travel quickly over sand? Explain your answer.

6 Explain why tractors need wide tyres to drive over muddy fields.

7 Explain why camels have large feet.

8 Ling is driving a car. When she stops it, she puts her foot on the brake pedal. She applies 10 N of force. The brake pedal has an area of 5 $cm^2$. Calculate the pressure Ling exerts on the pedal in $N/m^2$.

**Key facts:**

- ✔ To calculate pressure, use pressure = force ÷ area.
- ✔ The same force applied to a larger area produces a lower pressure.
- ✔ A larger force applied to the same area produces a higher pressure.
- ✔ The units of pressure are $N/m^2$.

## Check your skills progress:

- I can present my results in a table with the correct headings and their units of measurement.
- I can evaluate my experiment and suggest improvements, explaining any proposed changes.
- I can use symbols and formulae to calculate pressure.

Chapter 8 . Topic 4

# Pressure and diffusion in gases and liquids

You will learn:

- How to explain pressure in gases and liquids
- To describe the diffusion of gases and liquids
- To explain an analogy and how to use it as a model
- To use analogies
- To describe how evidence affects scientific hypotheses
- To make predictions based on scientific knowledge and understanding
- To plan investigations, including fair tests, while considering variables
- To choose experimental equipment and use it correctly
- To collect and record observations and measurements appropriately
- To present and interpret scientific enquiries correctly
- To reach conclusions studying results and explain their limitations

## Starting point

| You should know that... | You should be able to... |
| --- | --- |
| The particle model tells us that the particles in gases and liquids are moving around | Describe where a model works, and where it does not work |
| When particles gain energy they move faster and further apart | Plan an investigation to test an idea |
| | Identify control variables in investigations |

## Gas pressure

The particles in a gas are far apart and moving around. They collide with each other and the sides of the container they are in.

**8.25** *When you blow into a balloon you are putting more air into it. The idea of gas pressure helps explain why the balloon gets bigger.*

It is difficult to squash (compress) an inflated balloon.

This is because the gas particles inside are hitting the sides of the balloon. The force of the gas particles on the inside of the balloon is called **gas pressure**.

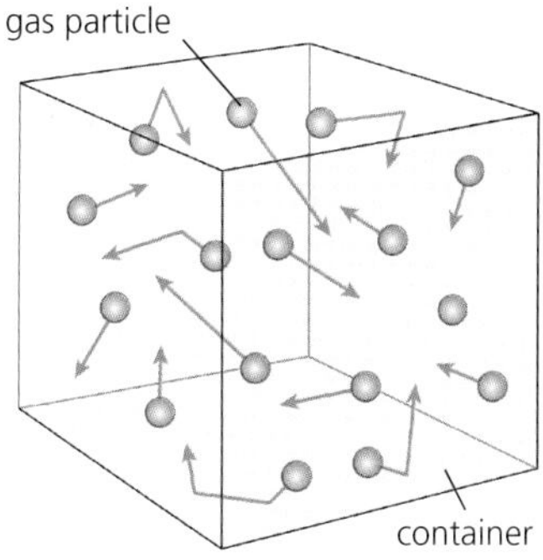


**8.26** *When gas particles hit a wall of their container there is a force on the wall. This force is the pressure of the gas.*

### Key term

**gas pressure**: the effect of the forces caused by collisions from gas particles on the walls of a container.

## Activity 8.4: Using the particle model to explain gas pressure

How can you make a model to show what causes pressure in a gas?

Your teacher will give you a container with some small balls in it. This is a model, or analogy, and the balls represent gas particles. The balls are free to move, like the particles in a gas.

**A** Shake the container so the balls move around.

They will hit each other and the sides of the container. These are collisions.

**A1**

**a)** State which part of the model shows the force of particles in a gas colliding with the sides of the container.

**b)** Use what you observed with the model to explain what causes gas pressure.

**B** Use the model to investigate what happens to gas pressure when:

**a)** The number of gas particles in the container changes

**b)** The size (volume) of the container changes.

**c)** The temperature of the gas increases.

**A2** Describe what you changed about the model to show:

**a)** A change in the number of gas particles in the container.

**b)** A change in the size of the container.

**c)** An increase in the temperature of the gas.

**A3**

**a)** Describe what happened to the 'particles' when you modelled an increase in temperature.

**b)** Explain what your model shows about the particles in a gas when the temperature is increased. Include a description of the energy transfers taking place in the gas.

Gas pressure is increased when there are more particles inside the container. This increases the frequency of collisions with the sides (how often collisions happen).

Increasing the temperature also increases gas pressure. This is because the particles have more energy, so move about more quickly. They collide with the walls of their container more often and with more force.

**1** Leo inflates a balloon.

**a)** Predict how the gas pressure inside the balloon changes as it is inflated.

**b)** Use the particle model to explain why.

**2** A vacuum contains no particles. If a balloon filled with air is placed into a vacuum, it starts to expand. Use your understanding of the particle model and gas pressure to explain this.

**3** Yasmin notices that her soccer ball feels much harder in the day, when it is hot, compared to at night, when it is cooler. Use the particle model to explain why.

## Pressure in liquids

Pressure in liquids is caused by the particles in the liquid moving around randomly and pushing against a surface. A boat floating in water experiences pressure from the liquid on all sides that are under the surface of the water. There is pressure on all sides because the particles that make up the water can move in any direction. The pressure is the total force of the liquid on the surface divided by the area of surface over which the force acts.

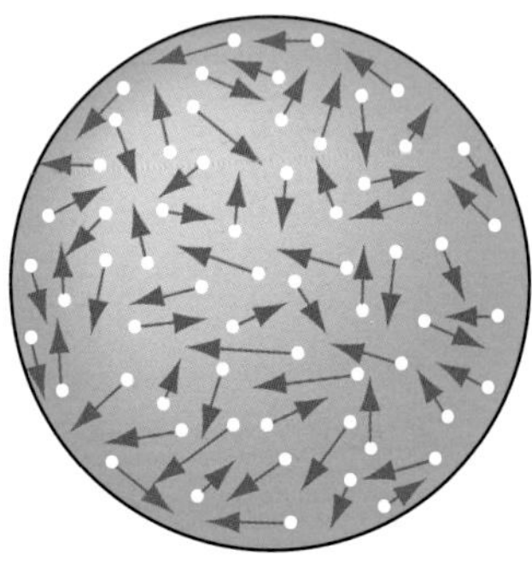

**8.27** *Just as with particles in a gas, particle movement in a liquid causes collisions. These collisions cause pressure on the sides of the container and on any surface in contact with the liquid.*

### Activity 8.5: Observing the effect of depth on pressure in a liquid

How does depth affect pressure in a liquid?

**A** Put some water in a plastic bottle. Make holes in the side of the bottle at different heights.

**B** Observe how far the water leaving each hole travels. You should notice that the water from the bottom hole travels a greater distance than the water from the top hole.

**A1** Where in the bottle is the highest pressure?

**A2** Where in the bottle is the lowest pressure?

**A3** Explain, using the idea of pressure, why the water in the bottom hole was pushed out furthest.

**8.28** *This shows that the pressure in liquids changes as depth increases.*

## Using liquid pressure

**8.29** *A hydraulic lift makes use of liquid pressure to lift heavy objects.*

Hydraulic devices use the pressure in a liquid to carry forces from one place to another. Figure 8.29 shows a hydraulic lift for a car. A person pushes down on the piston. This moves the liquid through the system. The pressure in the liquid under the car platform is the same as the pressure in the liquid under the piston. The piston has a smaller area than the platform. This means that the force of the liquid pushing up on the platform is bigger than the force used to push down the piston. This makes it easier for a person to lift the car.

4 **a)** The window on a submarine has an area of $1.50\,m^2$. The force of water on the hatch at a depth of 100 m is approximately 120 000 N. Calculate the pressure on the hatch.

**b)** The Jiah submarine in South Korea takes tourists underwater to 40 m. The submarine has small side windows to view the seafloor. Explain why it would be dangerous to increase the size of the windows.

## Air pressure

The Earth's atmosphere is a mixture of different gases including nitrogen, oxygen and carbon dioxide. All these particles are moving. Some collide with the surface of the Earth; other particles collide with the surface of any other object within the air. These collisions cause pressure, which we call air pressure (or atmospheric pressure).

### Changes of air pressure with height

The pressure of the Earth's atmosphere changes with height. Remember that all particles in the air have mass. Think of a mountain climber who starts their day at sea level. As they climb higher, there are fewer particles pushing down on them, so the mass of air above them is less, and the air pressure decreases.

**5** The atmosphere of Earth is sometimes described as like an 'ocean of air'. Explain how this comparison helps explain why atmospheric pressure is higher at sea level than at the top of a high mountain.

## Diffusion

Particles spread out from a region where there are lots of them (a high **concentration**) to a region where there are fewer (a lower concentration). This is called **diffusion**.

Diffusion happens in gases. If someone opens a bottle of perfume in a room, the perfume particles will diffuse through the air. The perfume particles spread out and intermingle with the gas particles in the air. The perfume particles spread out and intermingle with the air particles. They move from an area of high concentration where the bottle of perfume was opened, to the areas of low concentration. This process of diffusion continues until the perfume particles have spread out across the whole room and everyone in the room can smell the perfume.

Diffusion can also happen in liquids, because the particles in liquids can move around each other and intermingle. If you put a drop of red dye into a glass of water, after a few hours the dye particles will have spread out and intermingled with the particles of water. The dye colour spreads slowly through the water.

### Key terms

**concentration**: a measurement of how many particles of a certain type there are in a volume of liquid or gas.

**diffusion**: the spreading out of particles from where there are many (high concentration) to where there are fewer (lower concentration).

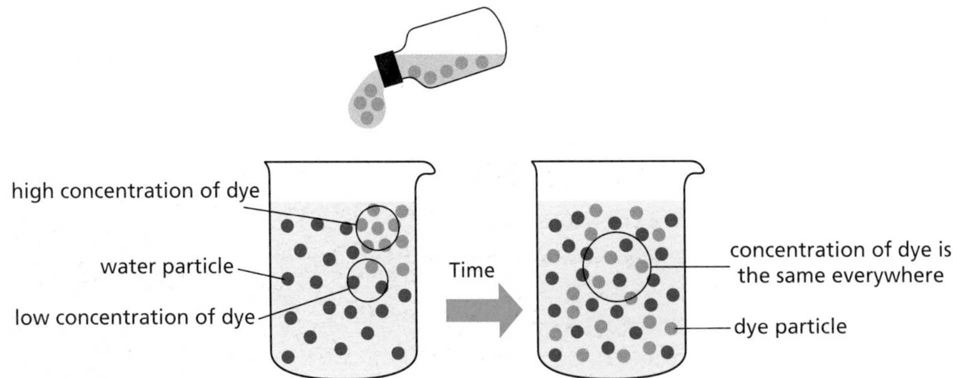


**8.30** *In the drop of dye, the dye particles are at a high concentration. The dye particles diffuse in the water until their concentration is the same throughout.*

## Activity 8.6: Using a model to observe diffusion

How can you use a model to show how diffusion occurs?

Your teacher will give you a tray with some small balls in it. The balls are different colours.

**A** Separate them, so the balls of one colour are at one end of the tray and the balls of the other colour are at the opposite end.

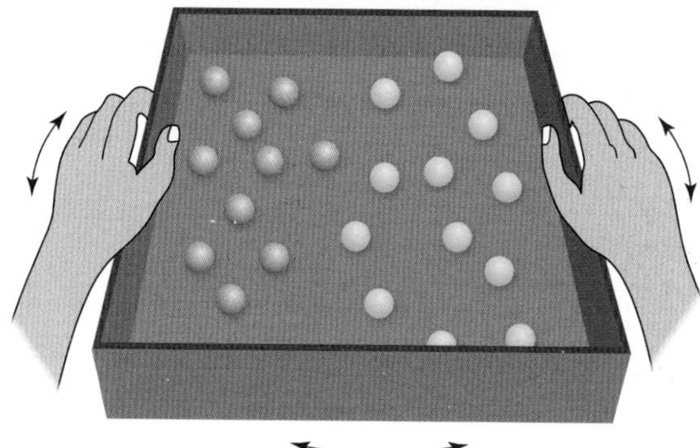

**8.31**

**B** Keep the tray on the table and shake it gently from side to side and up and down so the balls move around.

Questions to think about:

**A1** The balls represent the particles of gas. What does the tray represent?

**A2** What happens to the different coloured balls?

**A3** How does this model show diffusion?

**A4** Use the model to explain how diffusion happens.

Diffusion in a liquid or gas happens because particles are moving about randomly in all directions and collide with other particles.

When a collision happens, the particles move off in different directions. The particles intermingle. Eventually they will be evenly spread out.

## Uses of science: Evaluating issues

Factories, power plants and vehicles release polluting gases and particles of smoke into the air. Pollution affects people in the cities where it is produced, but it can also affect people outside cities and even in other countries, by processes that move particles through the air. The diffusion of pollutant gases such as carbon monoxide and nitrogen dioxide is very slow compared to the movement of a whole air mass by wind.

**8.32** *On days when there is no wind, air pollution in cities is much worse than on windy days. This is because diffusion is far too slow to move the polluting gases and particles away from where they are produced.*

## What affects the rate of diffusion?

The **rate** of diffusion is how quickly it happens.

Look at the experiment in Figure 8.33.

Both liquids evaporate and diffuse as gases through the tube. When the two gases meet they react to form a white solid.

You can see that the white solid does not form in the centre of the tube but closer to the hydrochloric acid. This is because the mass of each ammonia particle is less than the mass of each hydrochloric acid particle. They diffuse faster. The rate of diffusion is higher.

### Key term

**rate**: measurement of how quickly something happens.

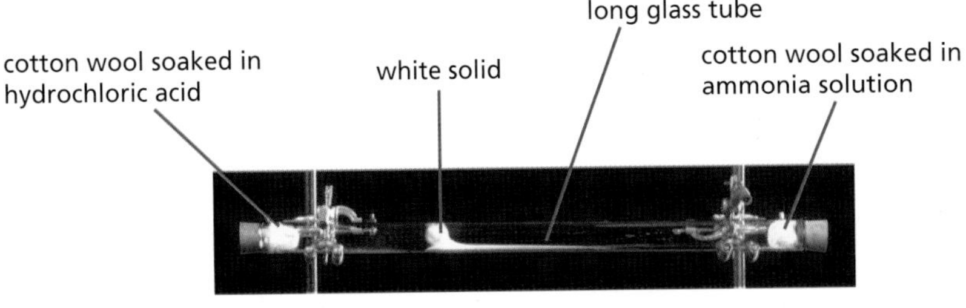


**8.33** *This experiment shows that mass of particles affects the rate of diffusion.*

## Activity 8.7: Making predictions about the rate of diffusion

How does temperature affect the rate of diffusion?

When a coloured sweet is added to water, the dye dissolves in the water and diffuses away from the sweet.

Use this to plan an investigation to investigate how the water temperature affects the rate of diffusion.

**A1** Make a prediction: how do you think temperature will affect how quickly diffusion happens? Use the particle model to give a reason for your prediction.

**A2** Write your plan:

- **a)** Which variable will you change? How will you change this and what range will you choose?
- **b)** Decide how you will compare the rates of diffusion. What variable, or variables will you measure?
- **c)** Are there any variables you should keep the same?

**A3** Write a list of all the equipment you will need.

**A4** Follow your plan and write your results in a table.

**A5** Write a conclusion. Say what the results show and compare them with your prediction.

**6** Use what you know about diffusion to explain why:

- **a)** You can smell what is cooking in the kitchen from another room.
- **b)** The colour from a teabag spreads throughout a cup of water.
- **c)** The smell of a scented candle is much stronger when it is lit.
- **d)** Diffusion happens more quickly in gases than in liquids.

### Making links

Topic 2.2 also applies the idea of particle movement, to explain the diffusion of gases between your lungs and your blood. Can you think of another example in living things where diffusion is important? What substances are involved?

## Key facts:

- ✔ Gas and liquid particles are always moving and so collide with each other and the sides of their container.
- ✔ The force of gas particles colliding with a surface is called gas pressure.
- ✔ The collisions of particles of a liquid with a surface causes pressure in a liquid.
- ✔ The particles in liquids and gases can spread out and intermingle with other liquids and gases in the process called diffusion.

## Check your skills progress:

- I can use the particle model to explain observations of pressure in gases and liquids, and diffusion.
- I can make and explain predictions based on scientific knowledge.
- I can plan an investigation and decide what equipment is needed.
- I can identify control variables in an investigation.

## End of chapter review

### Quick questions

1. Name the *two* forces which act on a boat floating motionless in the sea.
2. A car is driving at a steady speed along a straight road. What does this tell you about the forces on the car?
3. Two people pull on opposite ends of a rope.

   **(a)** The two people pull with the same force. What happens?

   **(b)** Person A pulls harder and person B continues to pull with the same force. What happens now?
4. Calculate the moments for the following situations in N m.

   **(a)** Akira sits on a see-saw. He sits 2 m from the pivot and weighs 50 N.

   **(b)** Nadia uses a screwdriver to open a tin of paint. The screwdriver is 0.3 m long and she applies 5 N of force.

   **(c)** Taksheel opens a door. The handle is 20 cm long and he applies 4 N of force.
5. Calculate the pressure produced on the ground in the following situations.

   **(a)** Mi wears shoes with an area of 0.05 $m^2$. She weighs 45 newtons.

   **(b)** Yura wears snowshoes (see the photograph on the right) with an area of 0.35 $m^2$. She weighs 70 newtons.

6. The spreading out of particles from a region where there of lots of them to a region where there are fewer is called...

   **(a)** gas pressure **(b)** diffusion **(c)** melting **(d)** condensation
7. Which *two* factors affect the rate of diffusion?

   **(a)** mass of the particles

   **(b)** speed of the particles

   **(c)** colour of the particles

   **(d)** shape of the particles

8. Which container would have the highest pressure?

   (a) A cold container with a large amount of gas.

   (b) A cold container with a small amount of gas.

   (c) A hot container with a large amount of gas.

   (d) A hot container with a small amount of gas.

9. Two weights, A and B, are placed on opposite ends of a pivot. In which of the following situations is it impossible for the moments to be balanced?

   (a) A weighs more than B and is closer to the pivot than B.

   (b) A weighs less than B and is further from the pivot than B.

   (c) A weighs more than B and is further from the pivot than B.

   (d) A and B weigh the same and they are the same distance from the pivot.

## Connect your understanding

10. Mandeep decided to investigate how much pressure he exerts on the ground when he is wearing different shoes.

    (a) Shoes do not have a regular shape. What could Mandeep do to work out their area?

    (b) What does Mandeep measure?

    (c) How does Mandeep use these measurements to calculate the pressure?

11. Dogs pull a sledge over a flat, icy surface. Their pulling force is 240 N forwards.

    Air resistance is 160 N and friction from the ice is 60 N.

    (a) Calculate the size and direction of the overall resultant force on the sledge.

    (b) Explain how you know whether the sledge is speeding up or slowing down.

12. Use particle theory to explain why you can compress a gas but not a liquid.

13. Rashin added a few drops of blue ink to a glass of cold water.

    (a) Describe what she will see in the glass. Explain why.

    (b) She then added a few drops of the ink to a glass of hot water. Describe how her observations would be different to what happened in the glass of cold water. Give a reason for this.

14. Sakura weighs 300 N and Abishek weighs 400 N. They both sit on the same side of a see-saw.

(a) Sakura sits 1.2 m from the pivot. Calculate Sakura's moment.

(b) Abishek sits 0.8 m from the pivot. Calculate Abishek's moment.

(c) Calculate Abishek's and Sakura's combined moment.

(d) Kyril sits on the see-saw on the opposite side of the pivot. Kyril weighs 500 N and sits 1.3 m from the pivot. Calculate Kyril's moment.

(e) State if the see-saw is balanced or unbalanced. Explain your answer.

15. Explain why snowshoes help people to walk over snow without sinking into it.

## Challenge questions

16. Cranes have counterweights. Use the diagram to explain why the counterweights must weigh more than the load.

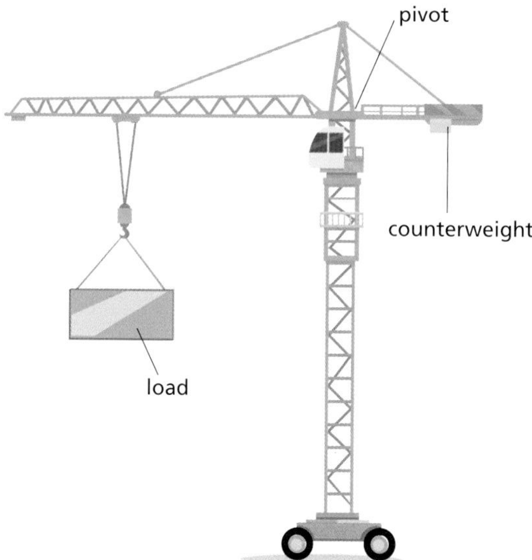


17. Yelena hits a nail with a hammer. The hammer head has an area of 12 cm$^2$. She hits it with a force of 24 N. Anton uses a hammer with an area of 20 cm$^2$. He hits the nail with a force of 36 N. Calculate the force Yelena and Anton hit the nail with and state who hits it with the greater pressure.

18. Newton's First Law of Motion predicts that, when there is no force acting on it, an object moves with a steady speed in a straight line. Explain why it is difficult to show this in an experiment. What might you try?

19. In humans and other mammals, air is breathed into the lungs.

Oxygen travels by diffusion from air in the lungs into the blood.
The blood must keep moving round the body to keep the rate of diffusion high.
Explain why.

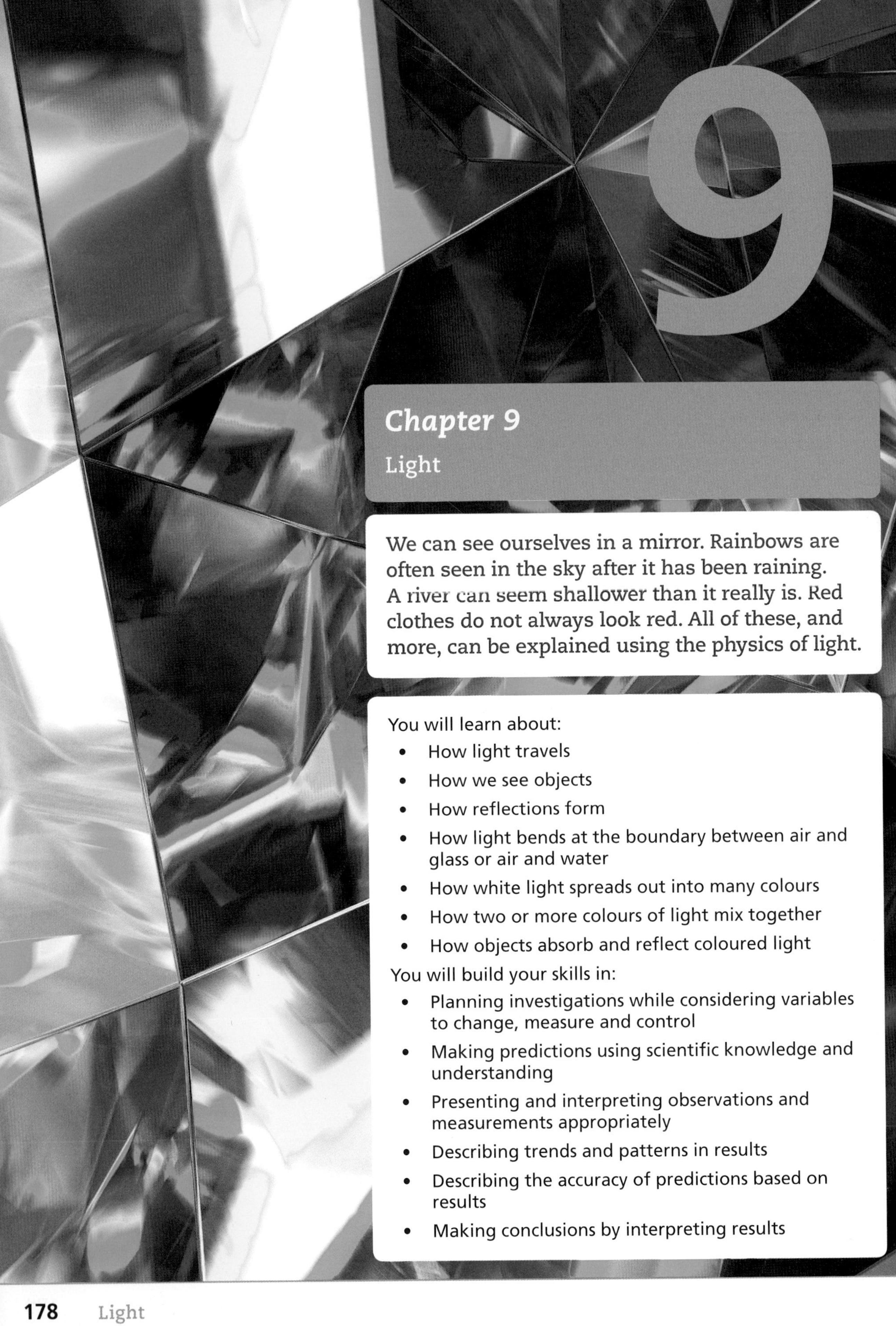

# Chapter 9

## Light

We can see ourselves in a mirror. Rainbows are often seen in the sky after it has been raining. A river can seem shallower than it really is. Red clothes do not always look red. All of these, and more, can be explained using the physics of light.

You will learn about:

- How light travels
- How we see objects
- How reflections form
- How light bends at the boundary between air and glass or air and water
- How white light spreads out into many colours
- How two or more colours of light mix together
- How objects absorb and reflect coloured light

You will build your skills in:

- Planning investigations while considering variables to change, measure and control
- Making predictions using scientific knowledge and understanding
- Presenting and interpreting observations and measurements appropriately
- Describing trends and patterns in results
- Describing the accuracy of predictions based on results
- Making conclusions by interpreting results

Chapter 9 . Topic 1

# Reflection

You will learn:

- To describe reflection at a plane surface
- To use the law of reflection
- To take accurate measurements and explain why this matters
- To collect and record observations and measurements appropriately
- To present and interpret scientific enquiries correctly
- To describe the application of science in society, industry and research
- To discuss the global environmental impact of science

## Starting point

| You should know that... | You should be able to... |
|---|---|
| Light travels in straight lines | Do experiments using light rays |
| A light ray is a straight line which shows the direction of light | Draw diagrams to show how light travels |

## How light reflects

If you look in a mirror you see yourself. This is called an image. This happens because light is reflected from the mirror's surface.

### Activity 9.1: Investigating how light reflects

Set up the equipment as in figure 9.1.

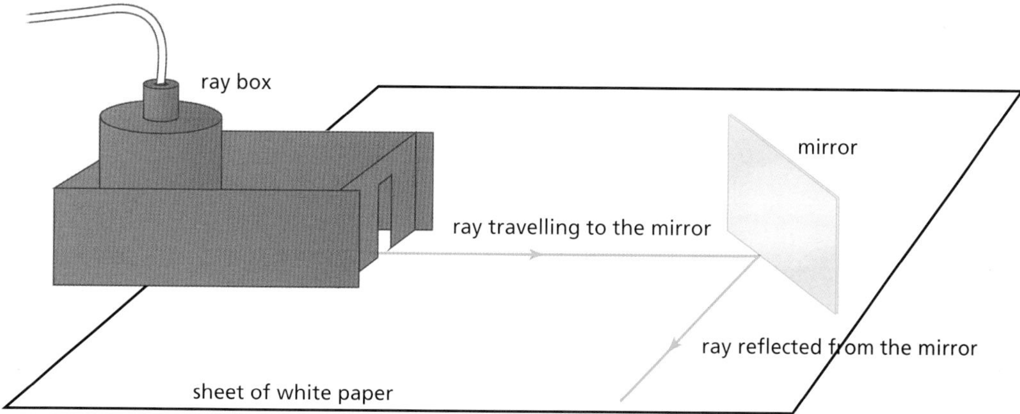


**9.1** *Investigating how light reflects.*

**A** Shine a ray of light from the ray box to the **plane mirror**. The ray box contains a light source. A slotted plastic filter is placed over the end which focuses the light from the light source to a single ray of light.

**B** Use a pencil to draw in the path of the ray towards the mirror (the **incident ray**) and after it reflects (the **reflected ray**). Add an arrow to each ray to show the direction in which it travels.

**C** Draw a dotted line at 90° to the mirror at the point where the ray hits the mirror (see figure 9.2). This line is called the **normal**.

**D** Measure the **angle of incidence** and the **angle of reflection** (see figure 9.2).

**E** Repeat for the following angles of incidence: 10°, 20°, 30°, 40°, 50°, 60°, 70°.

**A1** What do you notice about the angles of incidence and reflection?

Now plot your results in a graph. Plot angle of incidence on the x axis and angle of reflection on the y axis. Graphs help us to see the pattern in results. Drawing a **line of best fit** through the points makes the pattern clear. For this graph, your line of best fit should be a straight line.

**A2** Saira did a similar experiment and got the results shown in the table below.

| Angle of incidence ° | Angle of reflection ° |
|---|---|
| 0 | 0 |
| 15 | 15 |
| 30 | 30 |
| 45 | 40 |
| 60 | 60 |
| 75 | 75 |

**Table 9.1** *Angles of incidence and reflection when a ray of light is reflected.*

Plot these results on a graph with angle of incidence on the x-axis and angle of reflection on the y-axis.

Draw a straight line of best fit.

Do any of the points not quite fit on the straight line?

Odd results like this are described as **anomalous**. Perhaps somebody made a mistake when they made the measurement, so it is always a good idea to repeat your measurements.

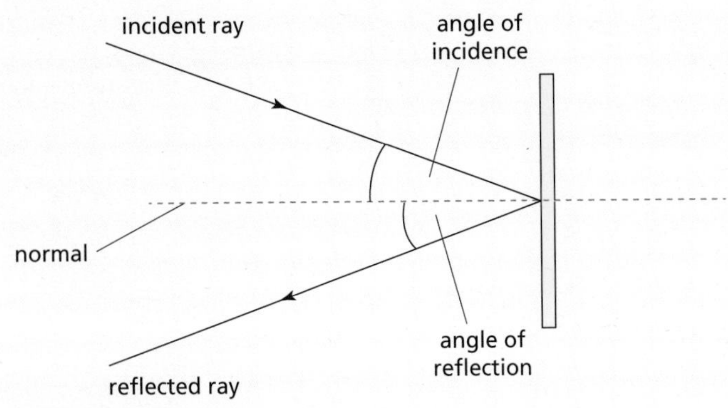


**9.2** *How light reflects.*

**Key term**

**anomalous results**: results which don't fit the pattern of the other results obtained.

## Law of reflection

The law of reflection is:

**angle of incidence = angle of reflection**

Remember that both angles must be measured between the ray and the normal.

This law is always true for all shapes of mirror and all angles of incidence.

Some people think that light always reflects through an angle of 90°. Is that correct?

Draw a diagram to explain your answer.

## Scattering

The light from an object reflected from a plane mirror with a smooth surface follows the law of reflection.

The light rays are reflected off the plane mirror at exactly the same angle as they went into the mirror.

If the light rays meet a rough or uneven surface, each of the rays will hit the surface at a different angle. Each ray still follows the law of reflection but there are lots of light rays all hitting the surface and reflecting off at different angles. The light is reflected in lots of different directions, causing an effect known as **scattering**.

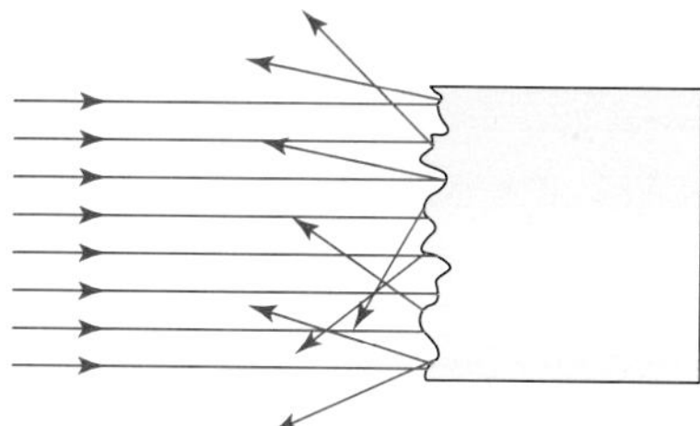

**9.3** *Scattering of light.*

## Periscopes

Periscopes let us see places we could not normally see. For example, submarines use periscopes so that they can see what is on the surface of the sea even when they are underwater. They use two mirrors.

Figure 9.4 shows a periscope being used to look over a wall.

### Key terms

**line of best fit**: a straight or curved line drawn through the middle of a set of points to show the pattern of data points.

**angle of incidence**: this is the angle between the incident ray and the normal.

**angle of reflection**: this is the angle between the reflected ray and the normal.

**incident ray**: this ray shows the light travelling towards the mirror.

**normal**: this is a line drawn at 90° to the mirror at the point where rays hit the mirror.

**plane mirror**: plane means flat, so a plane mirror is a flat mirror.

**reflected ray**: this shows the light travelling away from the mirror after it has been reflected.

**scattering**: scattering happens when light is reflected from particles and uneven surfaces.

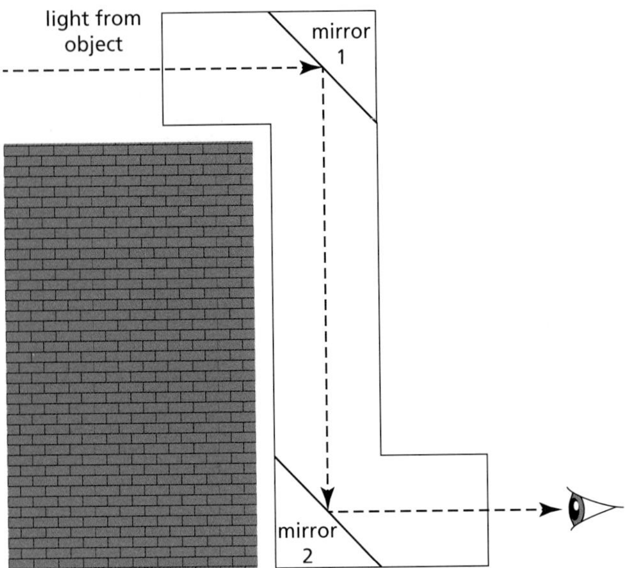


**9.4** *A periscope.*

## Activity 9.2: Making a periscope

Make a simple periscope using a cardboard tube or small box (e.g. a drink carton) and plastic mirrors.

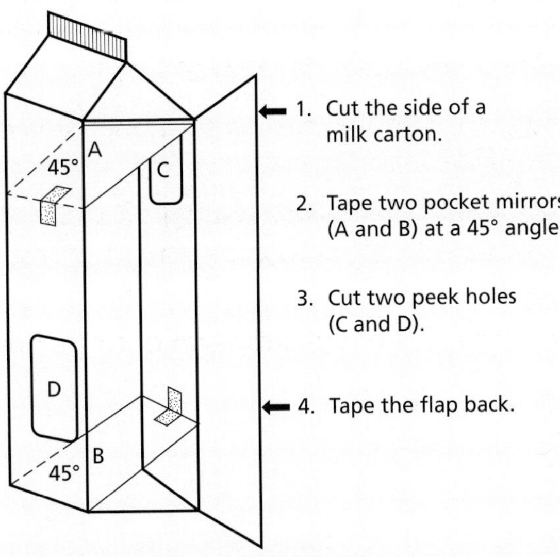


**9.5** *How to make a periscope.*

## Science in context: Mirrors and solar power

We can collect solar energy and store it using solar panels and batteries. Sometimes mirrors are also used to help concentrate the light from the Sun. This means that more solar power can be produced. Many hot countries generate a lot of their electricity this way. People who live in remote areas can generate their own electricity using solar power too. More and more homes are now generating at least some of their electricity using solar power. This helps reduce the use of fossil fuels. Doing this helps reduce pollution and global warming.

**9.6** *Large mirrors used in solar power plant.*

**9.7** *Many smaller mirrors used in solar power plant.*

## Making links

You have probably already studied echoes which are caused by sound reflecting.

Do you think that sound follows the same rules of reflection that light does?

## Key facts:

- ✔ The law of reflection states that the angle of incidence = the angle of reflection.
- ✔ The image in a mirror seems to be as far behind the mirror as the object is in front of it.
- ✔ A periscope uses two mirrors to reflect light.

## Check your skills progress:

- I can do an experiment to show the law of reflection.
- I can use diagrams to explain where the image of an object seen in a mirror seems to be.
- I can make a periscope and describe how it works.

# Refraction

You will learn:

- To describe refraction of light
- To take accurate measurements and explain why this matters
- To collect and record observations and measurements appropriately
- To present and interpret scientific enquiries correctly
- To describe the application of science in society, industry and research

## Starting point

| You should know that... | You should be able to... |
|---|---|
| Light travels in straight lines | Do experiments using light rays |
| When we draw light rays we add arrows to show the direction of the ray | Draw accurate diagrams to show how light travels |
| The normal is a line at 90° to a surface drawn at the point where a ray meets that surface | Use a protractor to measure angles accurately |
| Your brain sees an image where light waves entering your eye seem to have come from | |

## What is refraction?

Why does water look shallower from above than it really is? Why does a straight pencil look bent or broken when part of it is under water?

**9.8** *One effect of refraction.*

These things happen because of **refraction**. Refraction is the bending of light caused by a change in speed and direction of the light as it travels from one transparent substance into another.

The speed at which light can travel through a substance depends on whether the substance is a solid, liquid or a gas.

### Key term

**refraction**: the bending of light when it enters a different medium.

Light waves travel fastest through gases and slowest through solids. In a gas the particles are spread far apart so the light waves can pass through easily, so they travel faster. In a solid the particles are packed closely together so it is harder for the light waves to pass through and they travel slower.

When light waves travel from a material in one state to a material in a different state, their speed will change. Glass is a solid and air is a gas. When light waves pass from glass into air, their speed increases. When light waves pass from air into glass, their speed decreases.

Water is a liquid. When light waves pass from glass into water, their speed increases. When light waves pass from water into glass, their speed decreases.

The larger the change in speed, the larger the change in direction and the more the light will bend, or refract.

## Activity 9.3: Investigating refraction

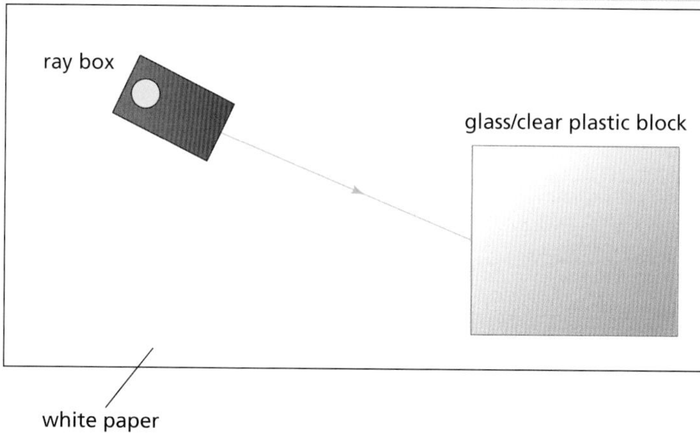


**9.9** *Investigating refraction.*

**A** Figure 9.9 shows a view of the equipment set-up used in this activity seen from above. Set up the equipment as shown in figure 9.9 in a darkened room.

**B** Draw around the glass block.

**C** Shine a ray of light into the block.

**D** Draw the path of the ray from the ray box to the block and after it leaves the block.

**E** Remove the block. Draw the path of the ray through the block using a straight line to join the places where it enters and leaves.

**F** Now measure the angle of incidence and the **angle of refraction**. (Note both angles are measured between the ray and the normal.) Put your result in a table.

**G** Replace the glass block in the same position as before. Move the ray box so the light enters the block at a different angle of incidence.

**H** Repeat for at least five different angles. Include using an angle of incidence of 0° (where the ray enters the block along the normal).

**A1** Look at your drawings and at figures 9.10 and 9.11. Choose the best word to complete these sentences:

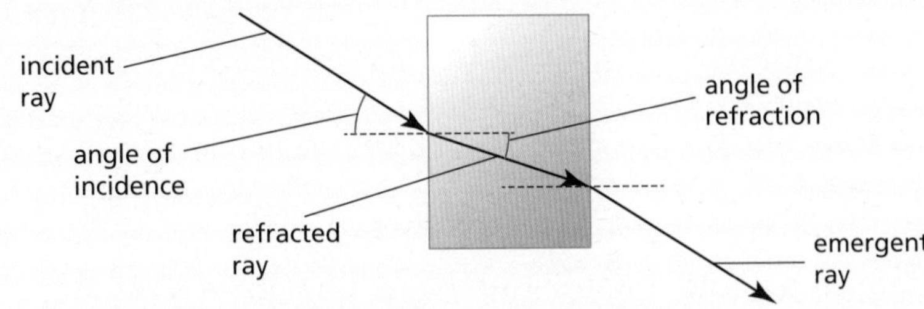


**9.10** *Refraction of light through a rectangular block*

**a)** Rays that enter the glass along the normal *bend/don't change direction*.

**b)** The ray bends *towards/away from* the normal when it enters the glass.

**c)** The ray bends *towards/away from* the normal when it leaves the glass.

**d)** The ray that leaves the glass is *parallel/not parallel* to the incident ray.

**e)** Unless the ray enters along the normal, the angle of refraction is always *larger/smaller* than the angle of incidence.

Now plot your results in a graph. Plot angle of incidence on the x-axis and angle of refraction on the y-axis. Draw a line of best fit through the points. For this graph, your line of best fit should be a curve. This graph shows that there is a link between the angle of incidence and the angle of refraction. The angle of refraction increases as the angle of incidence increases.

**9.11** *Refraction of light when the incident ray is along the normal.*

**A2** Do any of your points not quite fit on the line of best fit? If so, which results are anomalous?

## Real and apparent depth

Water looks shallower than it really is when viewed from above. Refraction explains this.

Imagine looking down into a stream like the one in figure 9.12. Light from a stone at the bottom travels through the water, into the air and into your eyes. When the light moves from the water into the air, it refracts.

When the refracted light enters your eyes, your brain thinks it has travelled in a straight line and comes from point X (the **apparent depth**). This is where your brain thinks the bottom of the stream (the **real depth**) is.

### Key terms

**angle of refraction**: the angle between the refracted ray and the normal.

**apparent depth**: how deep something appears to be.

**real depth**: how deep something really is.

**1** You look at a coin in a glass beaker. Its real depth is 10 cm. Is its apparent depth more than 10 cm or less than 10 cm?

**2** Light refracts more when it travels between glass and air than when it travels between water and air.

You look at a coin 20 cm under water. Its apparent depth is about 15 cm.

The same coin is under a glass block 20 cm thick. Is its apparent depth more than 15 cm, 15 cm or less than 15 cm?

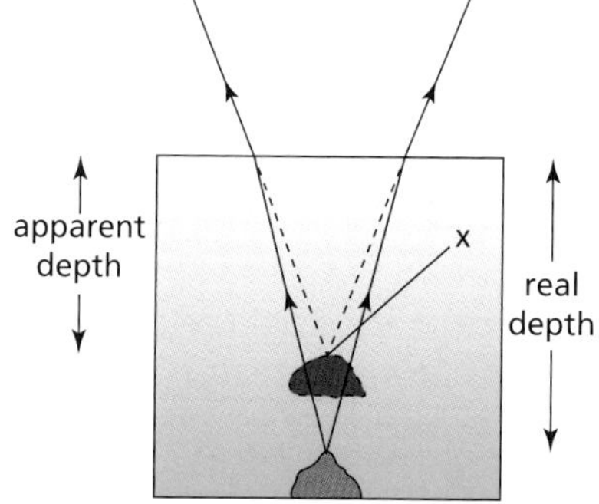


**9.12** *Real and apparent depth.*

Refraction can explain many optical illusions.

## Applying ideas

### Activity 9.4: The appearing coin trick

You will need a coin, a beaker and water.

**A** Put the coin into the beaker.

**B** Look at the coin, then move further away until you can't see the coin. Stay where you are.

**C** Ask your partner to gradually pour water into the beaker.

**D** Note when you can see the coin again.

**A1** Discuss why you can see the coin again when there is water on top of it.

### Activity 9.5: The bending pencil trick

You will need a pencil, ruler or straight piece of wood.

**A** Half fill a glass with water. Place the pencil in the water so it is half in and half out.

**B** Look at the pencil from above. Look at the pencil from the side. What do you notice?

**A1** Draw a diagram to show what you see from the side. Explain why this has happened.

## Science in context: Lenses and spectacles

Lenses use refraction. Light changes direction when it enters and leaves a lens. The earliest lenses were probably made from polished crystals. About 1000 years ago, an Arabic scientist, Ibn el-Haitam, gave the first known explanation of how lenses work. He also made a start at explaining how the eye works. These days most lenses are made of glass or Perspex. They can be made very accurately so that every lens refracts light exactly as we want it to. Spectacles use lenses. We do not know exactly who invented spectacles but it seems likely that they were invented in Italy about 750 years ago. Contact lenses are a much more recent invention.

**9.13** *Early pair of spectacles.*

## Key facts:

✔ Refraction is the bending of light when it travels from one transparent material into another.

✔ Light is refracted when it changes speed and therefore direction.

✔ Refraction makes water look shallower from above than it really is.

## Check your skills progress:

- I can do experiments to show how light refracts.
- I can use a ray diagram to show how light refracts through a rectangular glass block.

Chapter 9 . Topic 3

# Coloured light

You will learn:

- That white light is made of many colours
- To use a prism to show the dispersal of white light
- To describe what can be done with colours of light
- To reach conclusions studying results and explain their limitations
- To carry out practical work safely
- To describe the application of science in society, industry and research
- To describe how scientific progress is made through individuals and collaboration

## Starting point

| You should know that... | You should be able to... |
| --- | --- |
| Light refracts when it travels from air into glass, clear plastic or water | Draw accurate diagrams to show how light travels |

## Making a spectrum

If you shine white light into a **prism** (see figure 9.14), coloured light comes out at the other side. It looks like a rainbow. Why does this happen? Where do the colours come from?

White light is a mixture of different coloured lights. The prism splits the white light into these different colours. This is **dispersion**.

In figure 9.14, the light entering the prism is refracted when it goes into the prism and again when it leaves. White light is a mixture of all colours. As white light passes through the prism, different colours of light are refracted by different amounts. Figure 9.14 shows that violet light is refracted more than red light.

The colours in the light spectrum are:

Red Orange Yellow Green Blue Indigo Violet

You can remember the order of the colours from '**R**inse **o**ut **y**our **g**reasy **b**ottles **i**n **v**inegar'! You can make up your own saying to help you remember the colours too.

There are other ways of showing that white light is a mixture of the colours of the spectrum.

If you look at white light through a **diffraction grating**, you can also see a spectrum.

### Key terms

**dispersion**: the splitting of white light into a spectrum of colours.

**prism**: transparent object that refracts light.

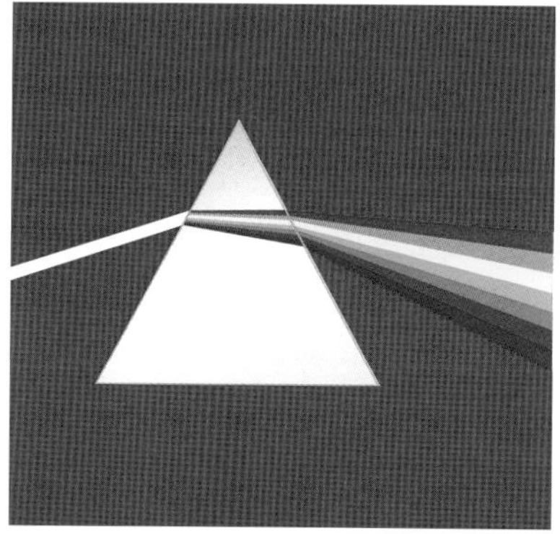

**9.14** *The spectrum formed by a prism.*

### Key term

**diffraction grating**: transparent piece of glass or plastic which has many lines drawn onto it. Light can pass through the spaces between the lines.

A diffraction grating is a transparent piece of glass or plastic. It has many lines drawn onto it. Light can pass through the spaces between the lines. There are usually about 3000 lines per centimetre. The closer together the lines are, the clearer the spectrum is. If the lines are too far apart, you won't see a spectrum at all.

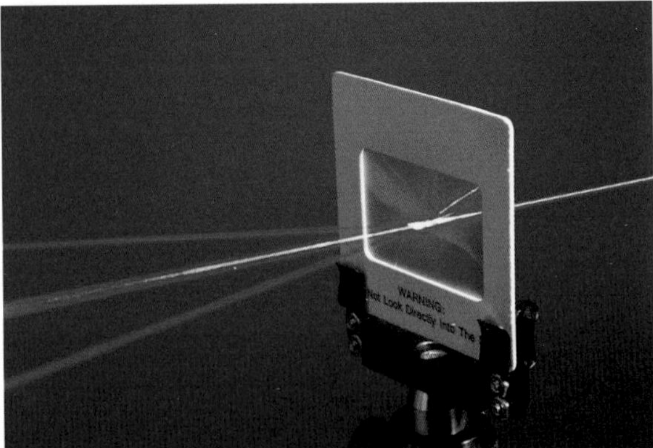

**9.15** *A diffraction grating.*

## Activity 9.6: Making a Newton's disc

**A** Cut a circle of card. Divide it into seven equal sections. Colour each section as shown in figure 9.16.

**B** Make a hole through the centre of your disc and put a pencil through the hole.

**C** Spin the disc as fast as you can. What do you see?

**A1** What does this experiment show?

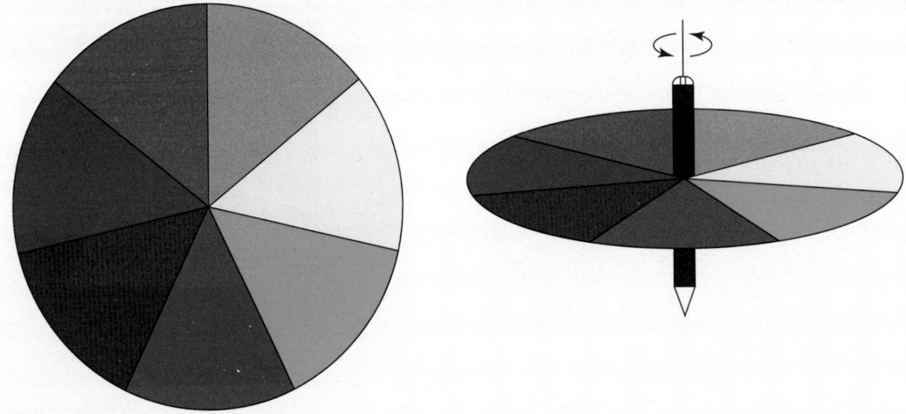

**9.16** *Newton's disc.*

## Combining colours

Look at figure 9.17. Why does the shirt on the right look blue and the shirt on the left look red? It is because the blue shirt only reflects blue light and the red shirt only reflects red light.

**9.17** *Coloured tops.*

## Primary and secondary colours of light

There are three **primary colours** of light:

Red Blue Green

These three colours can be added together to produce the secondary colours. This is called colour addition.

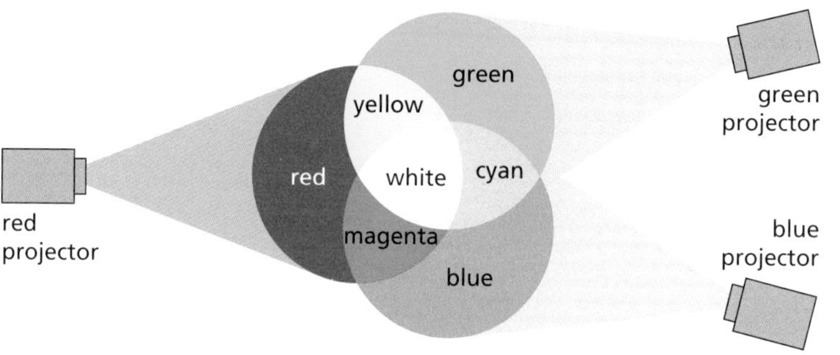


**9.18** *Primary and secondary colours.*

Yellow, magenta and cyan are called **secondary colours.**

Red and green light are added to form yellow light.

Red and blue light are added to form magenta light.

Green and blue light are added to form cyan light.

All three of the primary colours of light can be added together to produce white light.

Objects appear black when they do not reflect any light at all. We say that a black object absorbs all the light that shines on it. The effect is called **absorption**.

You can make many different colours by mixing different amounts of each primary colour. This is how you get coloured images on TVs and computer screens.

### Key terms

**absorption**: the way in which an object takes in the energy reaching its surface.

**primary colours**: red, blue and green. Mixing these colours of light together will make all other colours of light.

**secondary colours**: yellow, magenta and cyan.

## Colour filters and coloured lighting

A colour **filter** will only let light of its own colour pass through it. The filter absorbs all other colours of light.

For example, a red filter will only allow red light to pass through it.

Figure 9.19 shows what happens if you use different colours of light with different filters.

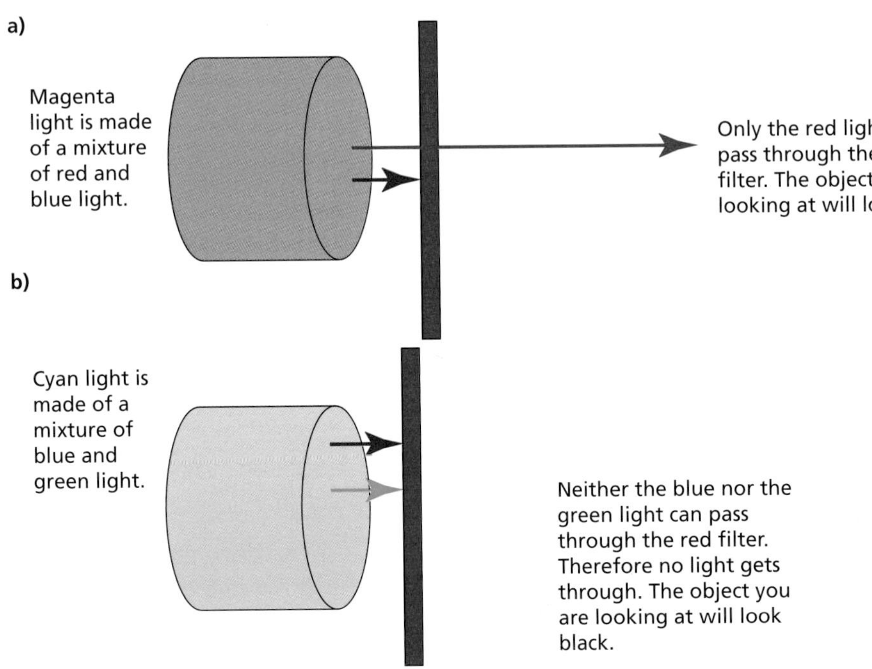


**9.19** *What happens when you use a) magenta and b) cyan light with a red filter.*

**1** What would you see if you looked at:

**a)** A yellow shirt using a red filter?

**b)** A magenta shirt using a blue filter?

**c)** A cyan shirt using a green filter?

**d)** A blue shirt using a red filter?

You can use colour filters to make different coloured lights. You can use different coloured lights to make objects look a different colour too.

In the process of colour subtraction one or more colours of light from white light may be absorbed by a material while the other colour or colours are reflected. The colours reflected are the colours seen.

White light is made up of red, blue and green light.

If white light is shone onto a shirt that absorbs red light, the red light is subtracted from the white light. The remaining

### Key term

**filter:** a colour filter will only allow light of its own colour to pass through it.

blue and green light would be reflected, and the shirt would therefore appear cyan to the observer due to the combination of the blue and green light.

This process of colour subtraction is further illustrated in figure 9.20.

Imagine you are looking at a sheet of white paper. White can reflect all the primary colours. Figure 9.20 shows you what happens when different colour lights are shone on different coloured papers.

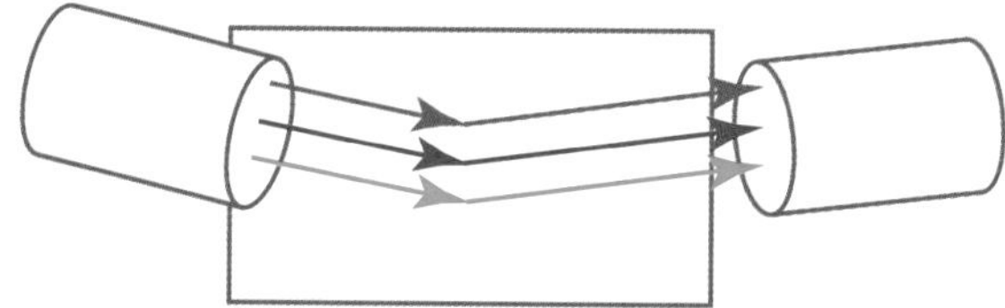

*White paper viewed under white light*: all three primary colours of light are reflected so the paper looks **white**.

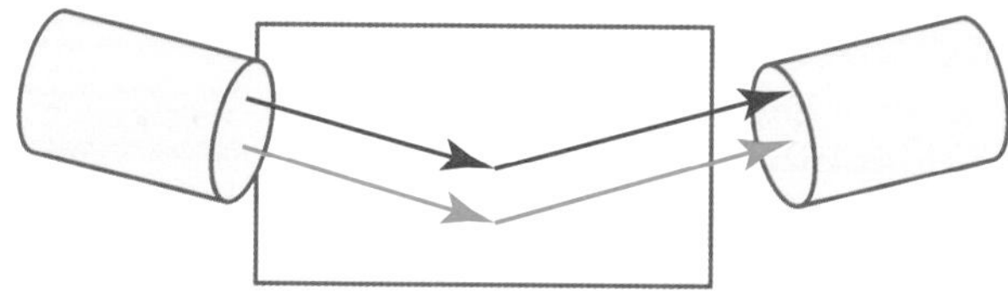

*White paper viewed under yellow light*: red and green light are reflected so the paper looks **yellow**.

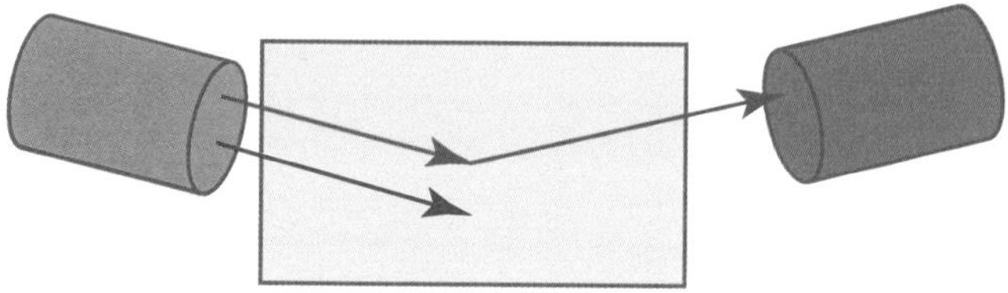

*Blue paper viewed under magenta light*: because the paper is blue, only blue light will be reflected so the paper looks **blue**. The red light is subtracted from the reflected light because the blue paper absorbs red light.

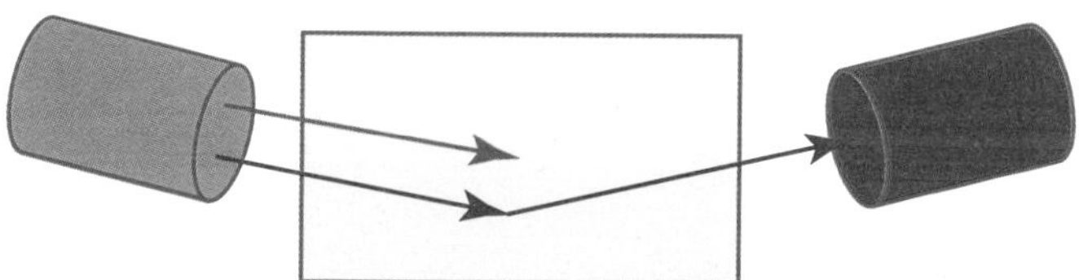

*Yellow paper viewed under magenta light*: yellow paper only reflects red and green light. The yellow paper absorbs the blue light. Blue light is subtracted from the reflected light. So, if you shine magenta light (a mixture of red and blue light) on it only red light will be reflected so the paper looks **red**.

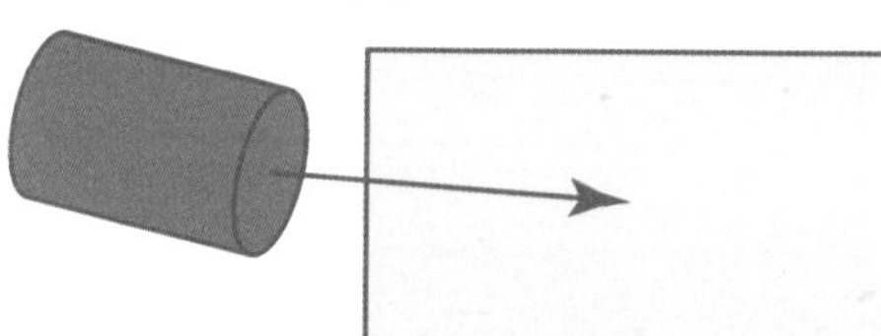

*Yellow paper viewed under blue light*: yellow cannot reflect blue light so the paper looks **black**.

**9.20** *Shining different coloured lights onto different coloured papers.*

2 What would you see if you looked at:

a) A magenta shirt under red light?

b) A red shirt under yellow light?

c) A cyan shirt under blue light?

d) A blue shirt using green light?

## Scattering

Almost everything scatters light. As we saw earlier in this chapter, when light hits a rough surface or particles in the atmosphere it reflects. Depending on the shape of the surface, the light will reflect in different directions.

Different colours of light are scattered by different amounts. Violet light is scattered the most and red light is scattered the least.

In the daytime, the sky looks blue because the Sun is high in the sky. Light from the Sun is scattered by the particles in the atmosphere. The angle of scattering means that more scattered blue light reaches your eyes. At sunrise and sunset, the Sun is lower in the sky. The angle of scattering means that more scattered red light reaches your eyes.

### Activity 9.7: Looking through a colour filter

**A1** Look at objects around you through different coloured light filters. What do you notice?

**A2** Use a red pen, a green pen and a blue pen to write on white paper. Look at the writing through a red filter, a green filter and a blue filter.

What do you see?

You could also try writing a secret message that can only be seen using a colour filter.

For example, when viewed through a red filter **WHESELXLLWCOOVMDE** says **WELCOME**.

### Activity 9.8: The effects of coloured lights

**A1** Use a ray box (or torch) and colour filters to make different coloured beams of light. Shine these onto different coloured objects and note any colour changes you see. Try to do as many colour combinations as you can.

**A2** You could also try writing a secret message that can only be seen using a coloured light.

For example, when viewed under green light **FHOSELXLLWDOB** says **HELLO**.

## Science in context: Colour blindness

The human eye contains cells called cones which detect colour. There are three different types of cone, each type detecting specific colours of light. If someone is colour blind it is usually because one or more type of cone does not work correctly. This means that the eye has problems detecting the difference between specific pairs of colours. The most common type of colour blindness is red-green where the colour blind person cannot tell the difference between these two colours though they have no problems with other colours.

Red-green colour blindness is usually diagnosed by using the Ishihara Plate test. This was developed by Dr Shinobu Ishihara in 1917. It is not as good at diagnosing other forms of colour blindness. Other tests have been developed more recently, for example the Farnsworth-Munsell 100 hue test which was developed in the 1940s. It was originally done using coloured tiles but is now usually done using blocks of colour on a computer screen.

Figure 9.21 shows three examples of the plates used in the Ishihara plate test. Inside each plate there is a number or a pattern created by one or more lines.

A person who is unable to see these patterns or numbers in the plates may have red-green colour blindness, and they should get tested further.

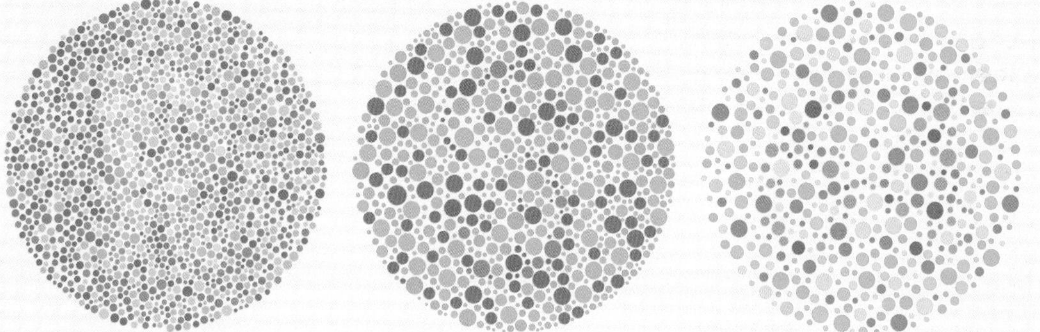

**9.21** *Examples of Ishihara Plates.*

It is important to diagnose colour blindness so that it does not affect children's learning and also because some occupations need perfect colour vision.

### Key facts:

- ✔ A prism can split white light into a spectrum. This is dispersion.
- ✔ The primary colours of light are red, blue and green.
- ✔ Adding primary colours can make secondary colours and white.
- ✔ Filters only allow certain colours of light to pass through them.
- ✔ Filters absorb other colours of light, causing them to be subtracted from the reflected light seen.

### Check your skills progress:

- I can do experiments to show how a prism can form a spectrum.
- I can use filters to make different coloured lights and make objects look a different colour.
- I can predict what something will look like under different colour lights and through different coloured filters.

## End of chapter review

### Quick questions

1. What do each of the letters A, B, C, D and E stand for in the diagram below?

   Choose from these words.

   | angle of incidence | angle of reflection | incident ray | normal | reflected ray |
   |---|---|---|---|---|

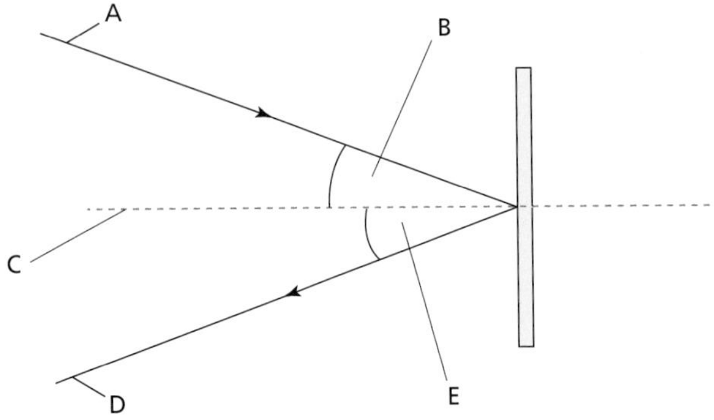


2. If you stand 60 cm in front of a plane mirror, how far behind the mirror will your image seem to be?

3. Which one of these statements about looking at yourself in a mirror is correct?

   **A** Your image is the right way up and the right way round.

   **B** Your image is the right way up and back to front.

   **C** Your image is upside down and the right way round.

   **D** Your image is upside down and back to front.

4. Put these colours in the order they appear in a spectrum, starting with red.

   orange green indigo yellow violet blue

5. Shoppers often take clothes outside to look at them in daylight before deciding whether to buy. Why do they do this?

6. What would the writing below say when viewed through a red filter?

   **JAREXEF DDP LUIMY GJ JHB CT**

## Connect your understanding

7. What would you see if you looked at:

   **(a)** A magenta T-shirt under blue light?

   **(b)** A red T-shirt under cyan light?

   **(c)** A cyan T-shirt under green light?

   **(d)** A magenta T- shirt under green light?

   **(e)** A blue T-shirt under cyan light?

8. Light is refracted when it travels from glass into air. Explain what is meant by the term 'refracted' and why this happens.

9. **(a)** Copy and complete the diagram to show where the fish appears to be.

   **(b)** Explain why the fish appears to be in the position you have drawn it.

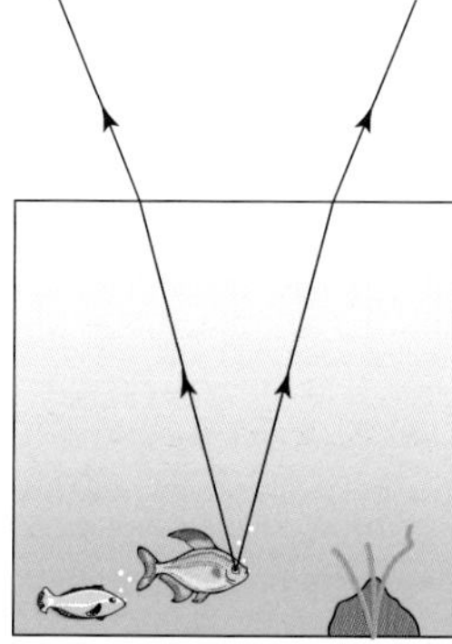

10. What would the tops shown below look like when viewed with:

    **(a)** green light

    **(b)** magenta light

    **(c)** cyan light

    **(d)** yellow light?

11. If you shine white light into a prism it produces a spectrum. This is one piece of evidence to support the idea that white light is made from a mixture of different coloured lights.

    Give *two* more pieces of evidence that show white light is made from a mixture of colours.

## Challenge questions

12. You can sometimes see a rainbow when it is raining. A water droplet can behave like a prism. Suggest how a rainbow is formed.

13. What would the writing below say when viewed through a yellow filter?

    J T H E A N O N D W K Y M O Q U C

14. Look at the arrangement of filters below. The first filter is blue and the second is magenta.

(a) What colour would you see if the source gave out:

- red light
- cyan light
- yellow light
- white light
- magenta light?

(b) Would it make any difference if the filters were the other way around?

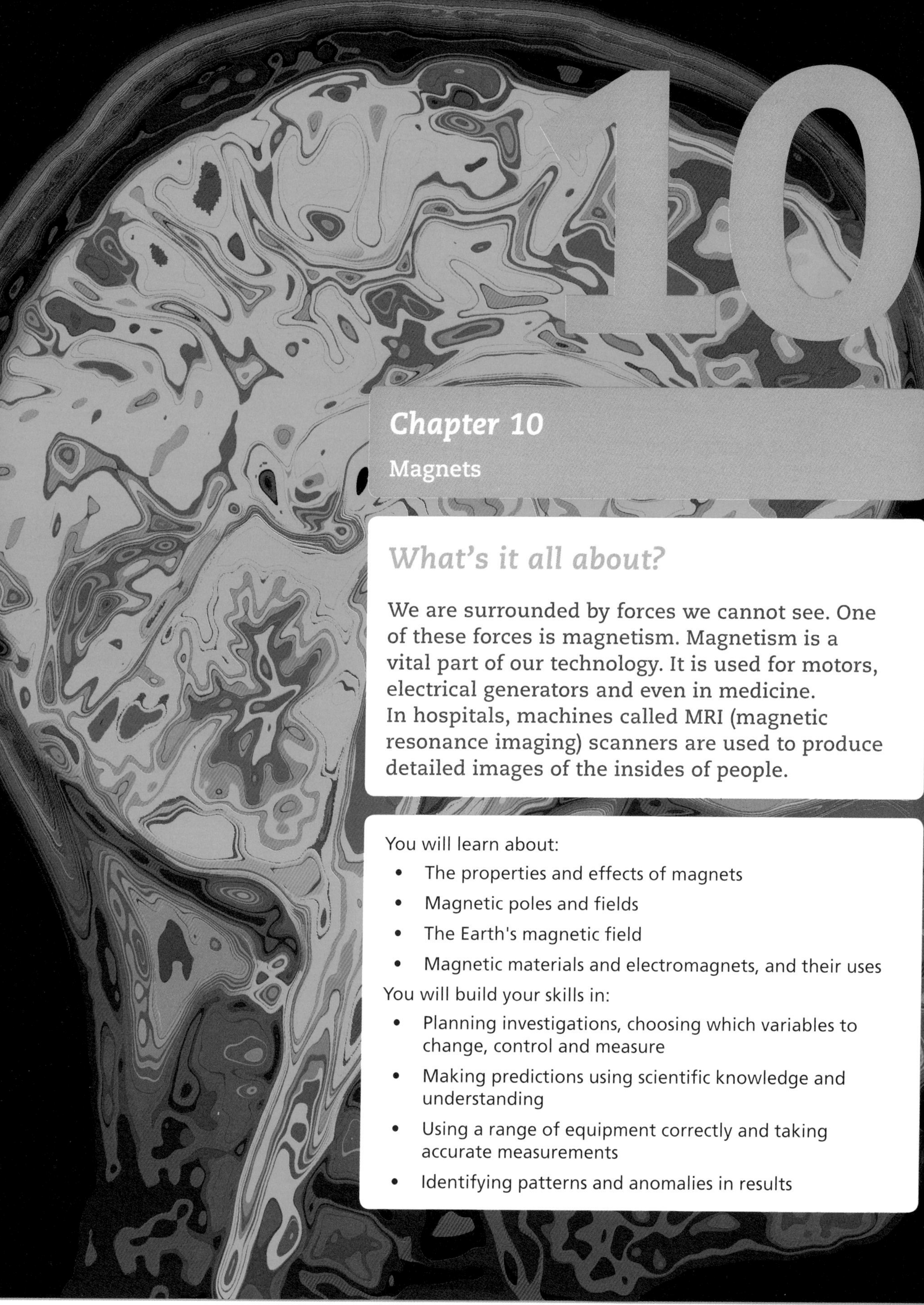


# Chapter 10

## Magnets

### What's it all about?

We are surrounded by forces we cannot see. One of these forces is magnetism. Magnetism is a vital part of our technology. It is used for motors, electrical generators and even in medicine.
In hospitals, machines called MRI (magnetic resonance imaging) scanners are used to produce detailed images of the insides of people.

You will learn about:

- The properties and effects of magnets
- Magnetic poles and fields
- The Earth's magnetic field
- Magnetic materials and electromagnets, and their uses

You will build your skills in:

- Planning investigations, choosing which variables to change, control and measure
- Making predictions using scientific knowledge and understanding
- Using a range of equipment correctly and taking accurate measurements
- Identifying patterns and anomalies in results

Chapter 10 . Topic 1

# Magnets and magnetic materials

You will learn:

- To describe a magnetic field
- That a magnetic field surrounds a magnet
- The force exerted by a magnetic field
- The reason why Earth has a magnetic field
- To describe how evidence affects scientific hypotheses
- To plan investigations, including fair tests, while considering variables
- To choose experimental equipment and use it correctly
- To carry out practical work safely
- To make predictions based on scientific knowledge and understanding
- To collect and record observations and measurements appropriately
- To present and interpret observations and measurements in an appropriate way
- To reach conclusions studying results and explain their limitations
- To discuss how scientific knowledge is developed

## Starting point

| You should know that... | You should be able to... |
|---|---|
| Forces change the speed and direction of objects | Outline plans to carry out investigations, considering the variables to control, change or observe |
| The weight of an object is due to the force of gravity | Make predictions and review them against evidence |
| The effects of a force can be measured even though the force itself cannot be 'seen' | Make careful observations including measurements |
| | Make predictions referring to previous scientific knowledge and understanding |

## What is magnetism?

If an object changes speed or direction, you already know that a force must be acting. A compass needle turns and eventually stops to point in a particular direction. This means a force must be acting on it.

**10.1** *A compass.*

We say that there is a **magnetic force** acting on the compass needle. This magnetic force is due to the Earth. The Earth itself is a giant **magnet**!

### Key terms

**magnet**: an object, usually made from iron, nickel or cobalt, that has its atoms aligned so that the object has a magnetic field.

**magnetic force**: force that occurs when a magnet attracts another object or repels another magnet.

## Key facts about magnetism

Over time, scientists have investigated **magnetism**. They have discovered that:

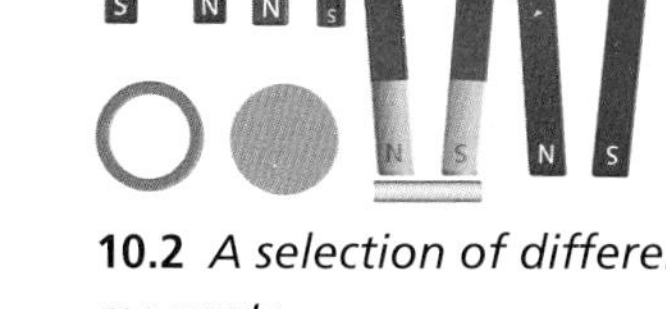


10.2 *A selection of different magnets.*

- Only some materials can be made into **permanent magnets**.
- These materials can be formed into magnets of different shapes, such as bar magnets or horseshoe magnets (see figure 10.2).
- All magnets have **magnetic poles** – a north pole and a south pole. (A north pole or a south pole cannot exist on its own.)
- If the north pole of one magnet is brought close to the south pole of another magnet, the poles attract (see figure 10.3). We say that 'unlike poles **attract**'.
- If the north pole of one magnet is brought close to the north pole of another magnet, the poles repel. We say that 'like poles **repel**'.

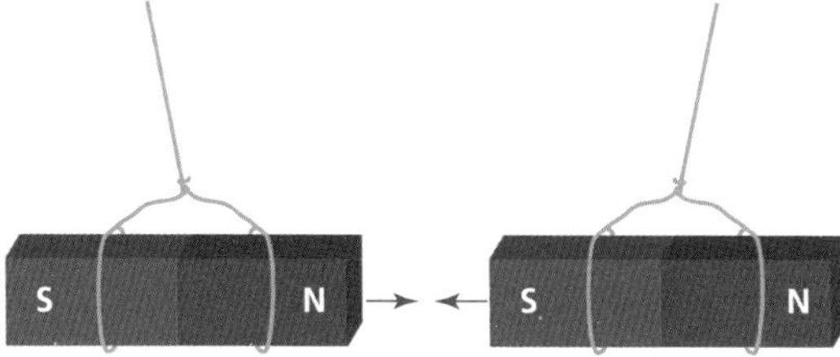


**10.3** *If two opposite poles are placed near each other, they attract.*

### Making links

Forces of attraction are also covered in Physics Chapter 8.1. Magnetic forces exert push and pull forces, and the pull forces of magnetic attraction can be compared with gravitational attraction.

State one way in which gravitational forces and magnetic forces are similar, and one way in which they are different.

### Key terms

**attract**: pull closer together.

**magnetic pole**: point on a magnet where the force is strongest.

**magnetism**: property of some materials that gives rise to forces between these materials and magnets.

**permanent magnet**: object made from a magnetic material that retains its magnetism for a very long time.

**repel**: push further apart.

## Activity 10.1: Attract or repel?

Magnets attract magnetic materials and other magnets, but they also repel other magnets. In this activity you will investigate when magnets attract and when they repel. You will need two bar magnets, a clamp and a piece of thread or string.

**A1** Suspend one of the bar magnets on a thread or string from the clamp.

**A2** Bring one end of the other bar magnet close to one end of the suspended bar magnet and note what happens. Do the magnets pull each other together, or do they push each other apart? Make a note of the poles of the magnets.

**A3** Now swap the magnet you are holding and bring the other end of the bar magnet close to one end of the suspended bar magnet. What happens now? Do the magnets pull each other together, or do they push each other apart? Make a note of the poles of the magnets.

**1** A compass is a small bar magnet which is used to help find directions. Why do you think it is important that you keep magnetic materials away from compasses?

## Magnetic materials

Some materials are affected by magnetic fields. We can show this by bringing a bar magnet close to the material. If it is a magnetic material, it will be attracted to the bar magnet and the bar magnet will be attracted to the magnetic material.

Any object made from a magnetic material can be made to behave like a magnet. Slowly stroke a magnetic material with one pole of a bar magnet several times, always starting at the same end (see figure 10.5). When you take the bar magnet away, the object will behave like a weak bar magnet. The effect does not last long.

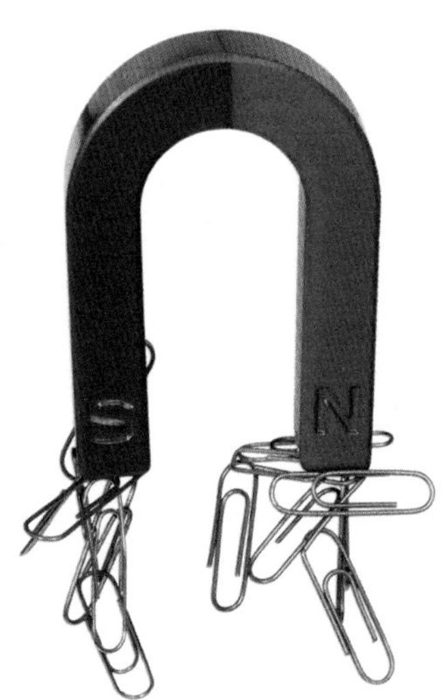


**10.4** *These paperclips are made of a magnetic material so they are attracted to the magnet.*

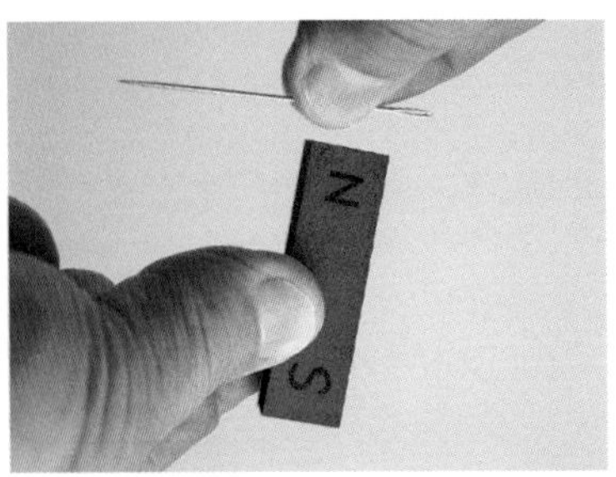


**10.5** *You can make a steel needle magnetic by stroking it with a magnet several times in the same direction.*

Magnetism is a property of the arrangement of the atoms in the structure of a material. We can shape magnetic materials, then **magnetise** the object using a very strong magnet.

### Key term

**magnetise**: to make magnetic.

### Activity 10.2: Magnetising a needle

You can use a magnet to magnetise a steel needle. After it has been magnetised, the needle will keep its magnetic properties for a while, but will gradually lose them.

**A1** Slowly stroke the needle with one pole of the magnet as shown in figure 10.5. To magnetise the needle properly, you will need to stroke it about 50 times. Make sure you stroke it in the same direction each time.

**A2** To test if your needle has been magnetised, see if it attracts one end of a compass needle.

Some materials are not affected at all by magnetic forces as you will find out in activity 10.3.

### Activity 10.3: Is this material magnetic?

Test some materials to see whether they are magnetic. You could try iron, paper, nickel, copper, aluminium, steel, card and plastic. Record your results in a table like this one.

| Magnetic materials | Non-magnetic materials |
|---|---|
| | |

**Table 10.1** *Magnetic and non-magnetic materials.*

**2** You have several different materials. Describe how you would test whether they are magnetic or non-magnetic.

## What is a magnetic field?

If you hold a paperclip near a permanent magnet, the material is attracted to the magnet. Even though you cannot see the force, you can feel its effect – you have to exert a force to hold the material in place.

The closer to the magnet you hold the magnetic material, the stronger the force gets. We say that there is a **magnetic field** surrounding the magnet. This field is stronger closer to the magnet, and weaker further away.

Key term

**magnetic field**: the region around a magnetic material in which a magnetic force acts.

## Seeing magnetic field lines

Iron filings are tiny flakes of iron. If you place a piece of white paper on top of any magnet and carefully scatter iron filings onto the paper, each filing lines itself up with the magnetic field of the magnet (see figure 10.6). Magnetic fields have the same general shape. Iron filings will reveal several lines that link from one pole to another.

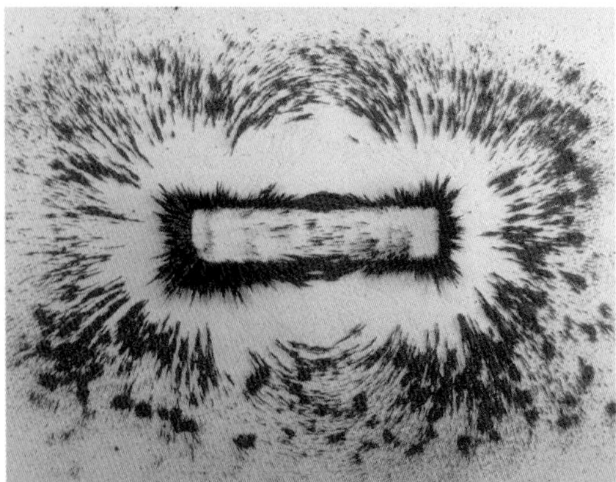

**10.6** *Iron filings revealing the magnetic field around a bar magnet.*

## Activity 10.4: Seeing the shape of a magnetic field

Magnetic fields are invisible, but this activity will let you see the shape the field forms around a bar magnet. You will need a piece of stiff white paper, a bar magnet and some iron filings.

**A1** Place the bar magnet on your work bench, and place the large sheet of paper on top of it so the magnet is roughly in the middle.

**A2** Carefully and slowly, sprinkle the iron filings over the paper around the magnet.

**A3** Gently tap the paper.

**A4** Sketch the pattern of the iron filings around the magnet.

When you have finished, take care putting the iron filings back in their container, and don't let them get stuck to the magnet.

Using iron filings in this way gives a good idea of the shape of the magnetic field, but it is not very precise. You can use a plotting compass, pencil and paper to produce a better diagram of the magnetic field (see figure 10.7). (A plotting compass is a small magnet that is suspended on a metal rod to turn freely.)

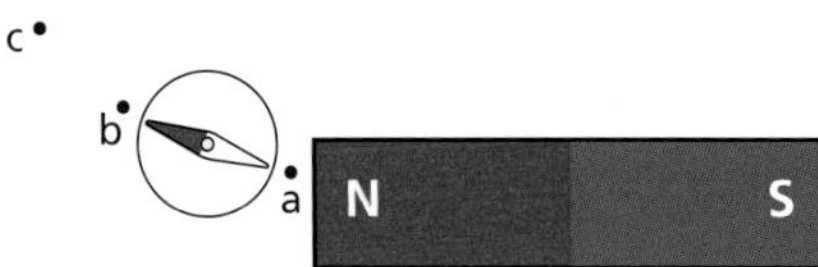


**10.7** *How to use a plotting compass to investigate the magnetic field around a bar magnet.*

## Activity 10.5: Drawing the shape of a magnetic field

This activity is similar to Activity 10.4 but you will use a small plotting compass instead of iron filings, so that you can draw the shape of the magnetic field directly onto your paper.

**A1** Place the bar magnet on top of a large sheet of paper and draw round the magnet.

**A2** Place the compass on the paper near the north pole of the magnet.

**A3** Mark a dot where the needle of the compass is pointing.

**A4** Then you move the compass so the back of the needle sits on the dot you have drawn. The needle may change direction. Mark a dot where the needle points.

**A5** Keep repeating this process all around the magnet until you get the field line.

## Drawing magnetic field lines

We can show this field by drawing magnetic field lines (see figure 10.8). The distance between the lines shows the strength of the field: the closer the lines, the stronger the field. The arrows on the field lines show which direction the plotting compass was pointing. Field lines go from north to south.

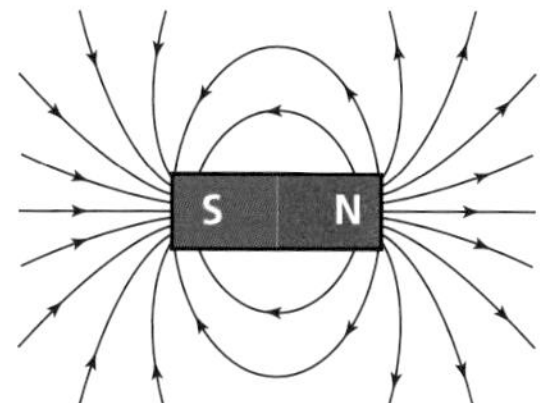


**10.8** *The field lines around a bar magnet.*

Remember these rules:

- Each line starts on a north pole and ends on a south pole.
- Each line should be smoothly curved and continuous (a field line must not break in the middle).
- Magnetic field lines cannot cross each other, join together or split apart.
- An arrow on each field line should go from the north pole to the south pole.

These rules apply to a magnet no matter what shape it is.

**3** **a)** The following magnetic field diagram contains at least four mistakes. Copy the diagram and explain each mistake.

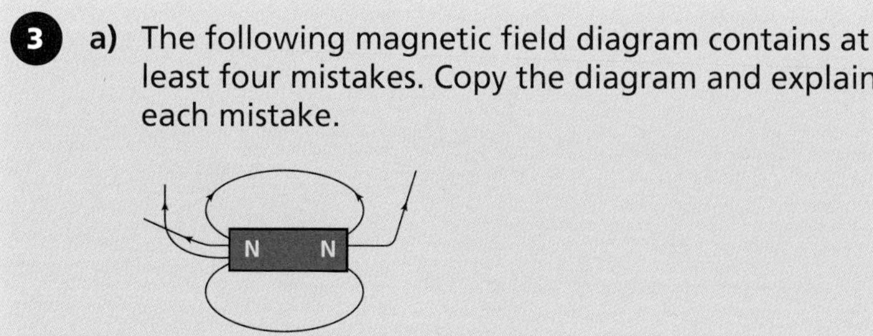


**10.9**

**b)** Draw a corrected diagram.

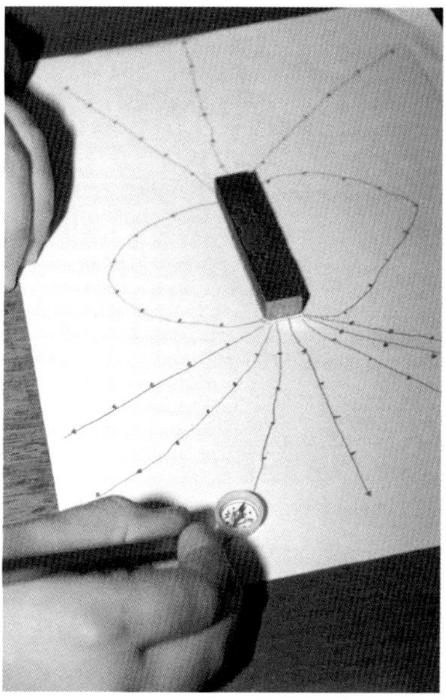

**10.10** *An experiment to find the field lines around a bar magnet.*

Figure 10.10 shows an experiment to find the field lines around a bar magnet.

## Field between two magnets

Remember:

- like poles repel
- magnetic field lines must not cross or break.

If you bring the north pole of one magnet to face the north pole of a second magnet, the two sets of magnetic field lines bend away from each other (see figure 10.11). The arrangement of field lines here shows that forces will act to push the magnets apart. The closer together the two north poles are, the closer the field lines and the stronger the forces.

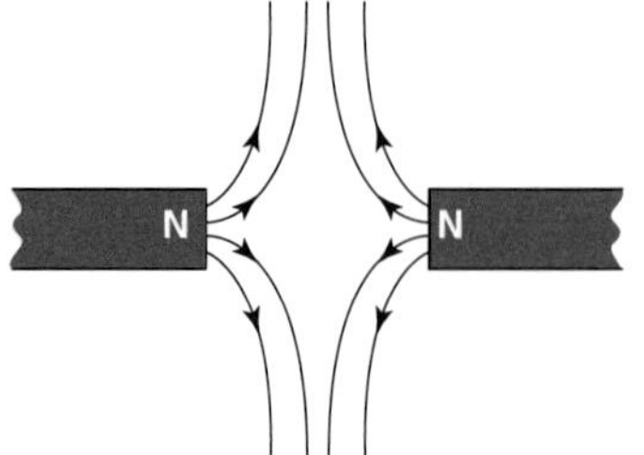


**10.11** *The field lines around two magnets that repel because like poles are facing each other.*

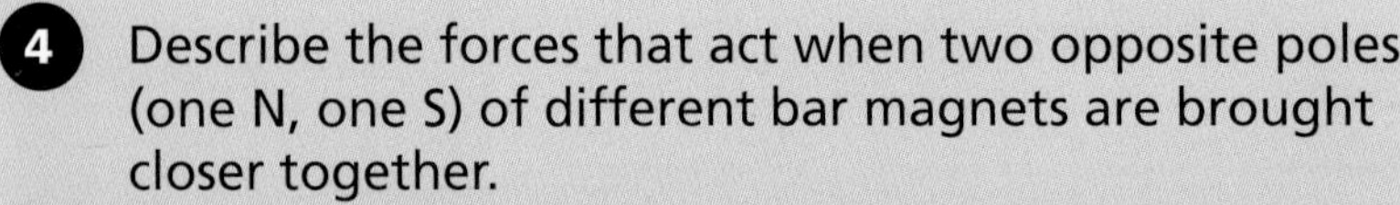

**4** Describe the forces that act when two opposite poles (one N, one S) of different bar magnets are brought closer together.

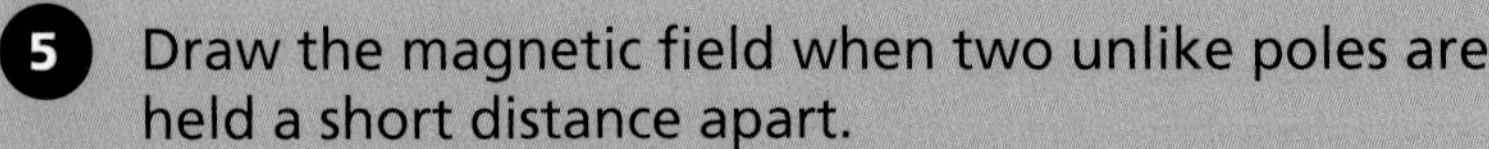

**5** Draw the magnetic field when two unlike poles are held a short distance apart.

**6** Use the field lines to predict what the magnets would do if allowed to move freely.

## The Earth's magnetic field

The Earth is a giant magnet. This means that if a magnet is left to move so that there is no friction on it, it will align itself with the Earth's magnetic field.

The magnet's north pole will point to the Earth's south pole and the magnet's south pole will point to the Earth's north pole. Since opposite poles attract, the area that compasses point to as the north pole is actually the magnetic south pole.

The Earth's magnetic field is what makes compasses point north and south. We can use this fact to help navigate, following compass bearings.

The Earth acts as a giant magnet because of its structure. The centre of the Earth, known as the core, is made up of iron and nickel, which are magnetic materials. The core is made up of the inner core and the outer core. The inner core is solid because it is under so much pressure, but the outer core is liquid because it is so hot.

As the Earth rotates on its axis, the liquid iron and nickel flow in whirlpools and set up electric currents, which in turn produce magnetic fields.

The magnetic field of the Earth is roughly (but not exactly) aligned with the axis of rotation of the Earth. The position of

### Making links

A compass is circular in shape, which means that our position can be measured in degrees. North is 000 degrees. Bearings are measured clockwise from north and given 3 digits. East is 090 degrees, south is 180 degrees and west is 270 degrees.

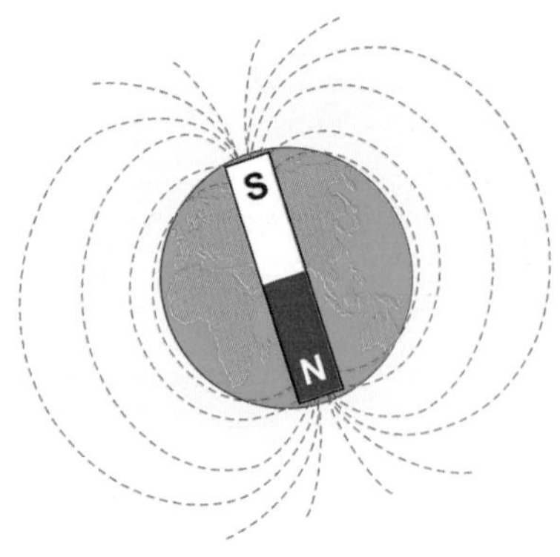


**10.12** *The Earth acts as a giant magnet.*

the magnetic north pole and the geographical north pole are not in the same place.

The magnetic north pole is the point on the Earth that compasses point towards. It is actually several hundred miles south of the geographic north pole and is always moving. In 2019, the magnetic north pole was calculated to be moving by about 50 km per year, heading from Canada to Siberia.

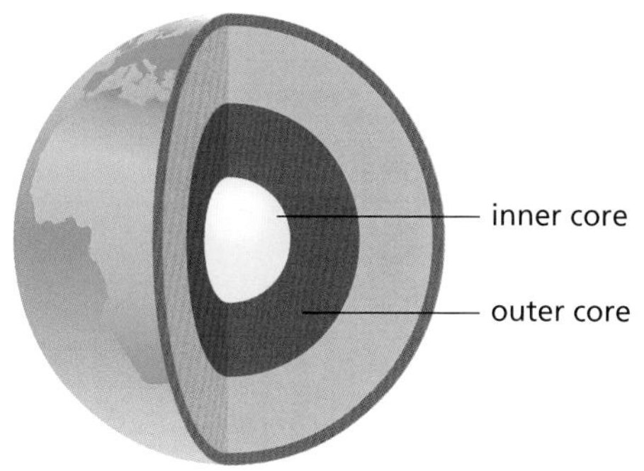


**10.13** *The Earth's outer core is liquid iron and nickel which creates a magnetic field.*

**7** Explain why all compasses point in the same direction.

**8** The north pole of a compass points towards the Earth's north pole. Use the idea of magnetic pole interactions to explain why this is actually the Earth's magnetic south pole.

## Science in context: The first compasses

At least 2400 years ago, ancient Greek and Chinese philosophers independently discovered a material that behaved strangely. If a piece of this material was shaped to form a pointer and allowed to spin freely, it would always come to rest pointing in the same direction. This happened no matter where the pointer was placed: indoors or outdoors, in the dark or in the light. This material is called lodestone.

**10.14** *An ancient Chinese compass made from lodestone.*

## Key facts:

- ✔ Magnetic materials experience magnetic forces.
- ✔ Every magnet has a north pole (N) and a south pole (S).
- ✔ Magnetic forces can attract or repel.
- ✔ Like poles repel, unlike poles attract.
- ✔ Magnetic field lines show the strength and direction of a magnetic field around a magnetised object.
- ✔ The Earth acts as a giant magnet because of the liquid iron and nickel in its core.

## Check your skills progress:

- I can investigate which materials are magnetic.
- I can magnetise an object using a permanent magnet.
- I can plot and draw magnetic field lines using a plotting compass.
- I can predict and explain the behaviour of two magnets when they are placed close together.

Chapter 10 . Topic 2

# Electromagnets

You will learn:

- To describe how to make an electromagnet
- That electromagnets have many applications
- To investigate what alters the strength of an electromagnet
- To describe how evidence affects scientific hypotheses
- To make predictions based on scientific knowledge and understanding
- To plan investigations, including fair tests, while considering variables
- To identify and control risks for practical work
- To choose experimental equipment and use it correctly
- To evaluate the reliability of measurements and observations
- To carry out practical work safely
- To collect and record observations and measurements appropriately
- To describe results in terms of any trends and patterns and identify any abnormal results
- To reach conclusions studying results and explain their limitations
- To present and interpret scientific enquiries correctly
- To describe the application of science in society, industry and research

## Starting point

| You should know that... | You should be able to... |
|---|---|
| Some metals have magnetic properties | Outline plans to carry out investigations, considering the variables to control, change or observe |
| Circuits are used to transfer electrical energy from one place to another | Make predictions referring to previous scientific knowledge and understanding |
| It is possible to measure the effect of a force even though the force cannot be seen | Make predictions and review them against evidence |
| | Choose appropriate apparatus and use it correctly |
| | Make careful observations including measurements |
| | Present results in the form of tables, bar charts or line graphs |
| | Make conclusions from collected data, including those presented in a table or graph |

## Electrical current creates a magnetic field

When a current flows through a wire, it creates a magnetic field. You can detect this magnetic field using plotting compasses or iron filings around the wire.

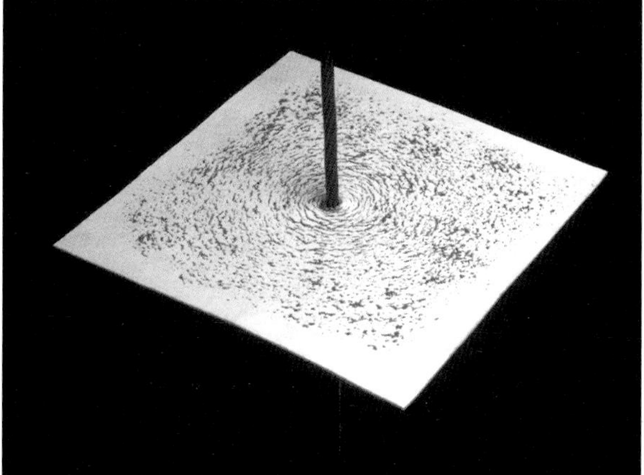

**10.15** *The magnetic field around a wire.*

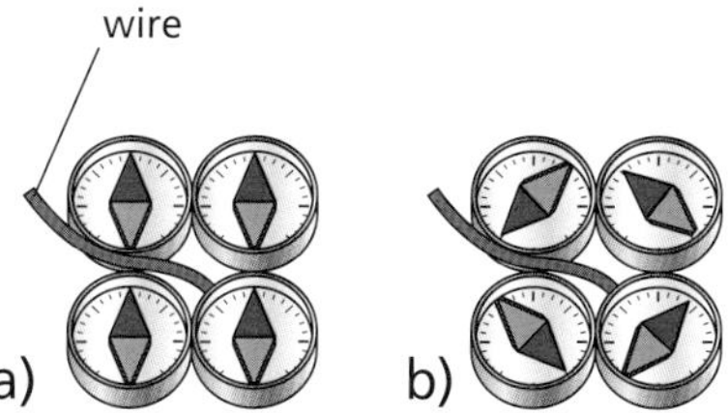


**10.16** *A plotting compass when a) the current in the wire is off and b) the current in the wire is on.*

## Using coils as magnets

If you coil a wire in a cylinder shape and pass a current through it, it creates a field that is similar to a bar magnet. We can coil this wire around a cylindrical piece of metal to increase its strength. We call this piece of metal a **core**. The material that a core is made from can affect the strength of the **electromagnet**.

Electromagnets are different to permanent magnets because you can turn an electromagnet on and off. You can also control the strength of an electromagnet. You can do this by changing:

- the number of turns in the coil: more turns create a stronger magnetic field
- the current flowing through the coil of wire: a bigger current creates a stronger magnetic field
- the metal the core is made from: a core made from a magnetic material such as iron creates a stronger magnetic field than a core made from a non-magnetic material, such as plastic.

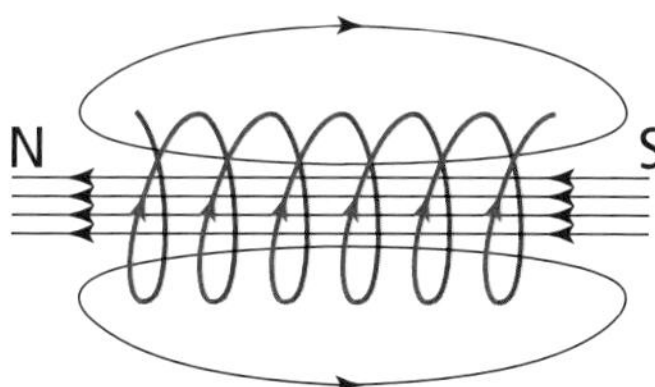


**10.17** *The magnetic field lines around a coil of wire.*

**10.18** *The wire is wrapped around a solid core.*

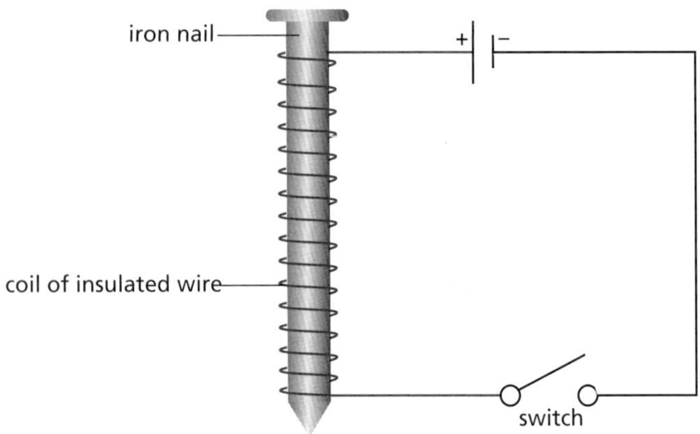

**10.19** *The circuit diagram for a simple electromagnet.*

## Key terms

**core**: piece of metal (usually iron) that a coil of wire is wound around to increase the strength of the magnetic field.

**electromagnet**: magnet that can be switched on or off using an electric current.

1. Describe how an electromagnet is different from a permanent magnet.
2. Tariq needs to choose a core for his electromagnet. He has iron, copper, plastic and wood. Which material would you recommend? Why?
3. You can change the strength of an electromagnet. Explain why this is useful.

## Activity 10.6: What affects the strength of an electromagnet?

In this activity, you will make an electromagnet and investigate the different factors that affect the strength of the electromagnet. The strength of an electromagnet depends on:

- the size of the current going through the wire
- the number of turns in the coil
- the type of core inside the coil of wire.

**A1** Construct the electromagnet: wind wire around a steel nail or bar. Put as many coils as you can around the bar without them overlapping. Insert connectors into a powerpack.

**A2** The steel bar will be your electromagnet. Test the strength of the electromagnet with no current. How many paperclips can you pick up with no current?

**A3** Gradually increase the current through the coils and test the strength of the electromagnet by counting the numbers of paperclips that can be lifted. Record your results in a table.

| Current (amps) | Number of paperclips |
|---|---|
| 0 | |
| 0.2 | |
| 0.4 | |
| 0.6 | |
| | |

**Table 10.2** *Number of paperclips that can be lifted by an electromagnet with different currents.*

**A4** Plot your results on a graph.

**A5** Turn off the current and increase the number of turns in the coil by adding a second layer of coils. Discuss with a partner what you think will happen when the current is turned on.

**A6** Repeat **A1–A4** with more coils of wire.

**A7** Turn off the current and make an electromagnet with a different core. Look back at Activity 10.3 to help you decide which materials to use. Put the same number of coils around the new core as you used in **A1**. Discuss with a partner what you think will happen when the current is turned on.

**A8** Test the strength of the electromagnet with the new core with no current. How many paperclips can you pick up with no current?

**A9** Gradually increase the current through the coils and test the strength of the electromagnet by counting the numbers of paperclips that can be lifted. Record your results in a table and compare the strength of the new electromagnet with the one you made in **A1**.

**A10** Discuss with your partner which factor made the biggest difference.

**4** Mira did an investigation on how the number of coils around an electromagnet affected the number of paperclips it picked up. She collected some results. Plot a graph of her results.

| Number of coils | Number of paperclips I picked up |
|---|---|
| 0 | 0 |
| 5 | 2 |
| 10 | 4 |
| 15 | 3 |
| 20 | 8 |
| 25 | 10 |

**Table 10.3** *Mira's results.*

**a)** One of the results is anomalous. Which one?

**b)** Draw a straight line going through as many points as possible. Ignore any points that are a long way from the straight line.

**c)** What conclusion can Mira draw from the results?

**5** Lee changed both the number of coils and the current when he did his investigation. When he increased them both, he picked up more paperclips. Why can't he conclude that the number of coils affects the strength of the electromagnet?

**6** Kyra decided to investigate how the material of the core affected the strength of the electromagnet.

**a)** How many independent variables should an investigation have? Explain your answer.

Table 10.4 shows Kyra's results.

| Core material | Number of paperclips I picked up |
|---|---|
| Plastic | 1 |
| Iron | 12 |
| Steel | 10 |
| Nickel | 8 |
| Aluminium | 2 |
| No core | 1 |

**Table 10.4** *Kyra's results.*

**b) i)** State which materials from the table are metals and which materials are non-metals.

**ii)** Kyra concluded that a metal core makes the electromagnet stronger. Is Kyra's conclusion correct? Explain your answer.

**iii)** Which metallic elements are the best at increasing the strength of a magnetic field?

**iv)** What is the link between the three metals that increase the strength of an electromagnetic field the most?

Key terms

**control variable**: variable you keep the same during an experiment.

**dependent variable**: variable you decide to measure in an experiment.

**independent variable**: variable you decide to change in an experiment.

## Uses of electromagnets

### Industrial lifting magnets

Large electromagnets are used in scrapyards to pick up and move large objects made from magnetic material, such as steel, and move them to other places. You can lift something up when the current is on, and drop it by turning the current off.

**10.20** *An industrial lifting magnet.*

## Electric bells

Electric bells use electromagnets. When you press the switch, you complete a circuit, and so turn on the electromagnet. This attracts the hammer to the bell which makes the noise. However, when the hammer strikes the bell, it breaks the circuit, which turns off the electromagnet. The hammer goes back to its original place, but this turns on the electromagnet again and attracts the hammer to the bell again. This continues until you stop pressing the switch.

**10.21** *An electric bell.*

## Generators

Wind turbines contain generators. A generator is a coil of wire that turns inside a magnetic field to generate a voltage.

When the wind blows the turbine blades spin and turn the coil of wire inside the magnetic field.

## MRI scanners

Magnetic resonance imaging (MRI) is a very useful tool in medicine. With MRI, the patient is put into a strong magnetic field created by an electromagnet. In this field, the atoms in the patient's body act like tiny magnets and align with the field. This can be used to produce an image of the inside of the body which can be used to detect many conditions such as tissue damage, diseases affecting the brain, some cancers, heart disease and diseases in bone joints. Since they only use a magnetic field, MRI scanners are much safer than other techniques such as X-ray scanning and you can see tissue as well as bone.

### Science in context: Maglev trains

Maglev trains use electromagnets to float above the rails. They are used in Japan, China and South Korea. Maglev is short for magnetic levitation. These trains have no moving parts and there is no friction between the train and the track. This means that they are faster and quieter than conventional trains. They also cheaper to maintain after they are built.

**10.22** *A maglev train.*

There are electromagnets on the train and the track. Some maglev trains have electromagnets that repel each other to make them levitate. Others have electromagnets that are attracted to electromagnets underneath the track and this makes them levitate.

7. Why can't you make an electric bell using a permanent magnet?

8. Maglev trains have very small maintenance costs. Give a reason why you think people haven't built more of them.

## Key facts:

- ✔ When a current flows through a wire, it creates a magnetic field.
- ✔ When a wire is coiled into a cylinder shape and a current is passed through it, it has a magnetic field like a bar magnet. This is called an electromagnet.
- ✔ The strength of an electromagnet is affected by the number of coils it has, the current and the type of core used to wrap the wires around.
- ✔ Electromagnets have many uses because they can be turned on or off and their strength can be changed.

## Check your skills progress:

- I can construct and use an electromagnet.
- I can use a range of equipment to discuss what affects the strength of an electromagnet.
- I can present my results in a table and graph.
- I can use my scientific understanding to discuss and explain what affects the strength of an electromagnet.

# End of chapter review

## Quick questions

1. Give definitions for the following terms:
   **(a)** permanent magnet
   **(b)** electromagnet
   **(c)** magnetic material
   **(d)** magnetic field.
2. There are three pieces of metal. One piece of metal is a permanent magnet. One piece of metal is a magnetic material. One piece of metal is a non-magnetic material. For each piece of metal, say if it is a permanent magnet, a magnetic material or not magnetic. Give a reason for your answer.
   **(a)** One metal is attracted to both poles of a magnet.
   **(b)** One metal does not move at all when a magnet is placed near it.
   **(c)** One metal is attracted to one end of a magnet and repelled by the other end.
3. Which of the following mixtures can be separated by a magnet and why?
   **(a)** Iron and nickel
   **(b)** Steel and plastic
   **(c)** Wood and copper
4. Give instructions for turning a steel needle into a permanent magnet.
5. Look at the diagrams below. Copy the diagrams and draw an arrow to show which direction the blue magnet will move in.

**(a)** 

**(b)** 

6. Copy the diagram below. Add arrows to the plotting compasses to show in which direction they point.

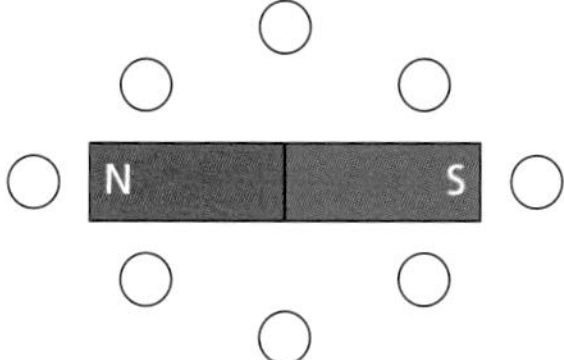

7. Describe a method, other than using plotting compasses, to see a magnetic field.

8. Copy the diagrams below. Draw fields around the magnets.

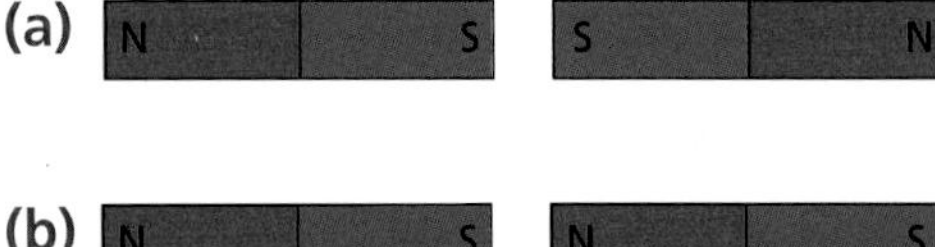


(b) N S N S

9. Which pole of a magnet points to the north pole of the Earth? Explain your answer.

## Connect your understanding

10. Omar investigated how the number of coils around an iron core affects the strength of an electromagnet. He measured the strength by measuring how many paperclips it picked up. He obtained the results given in the table.

| Number of coils | Number of paperclips picked up |
|---|---|
| 0 | 0 |
| 10 | 3 |
| 20 | 12 |
| 30 | 19 |
| 40 | 25 |
| 50 | 30 |
| 60 | 34 |

Plot a graph of his results and describe any trend or pattern.

11. Here are magnetic fields around two pairs of magnets.

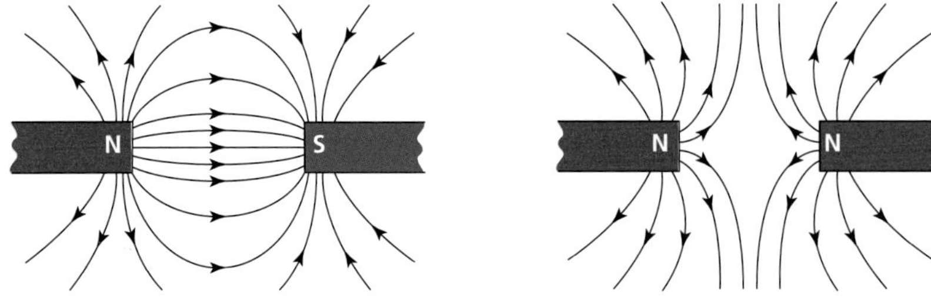


(a) Which two magnets are attracted?

(b) Which two magnets are repelled?

(c) Draw the field around two magnets with south poles facing each other.

## Challenge question

**12.** Use the diagram below to explain how an electric bell works.

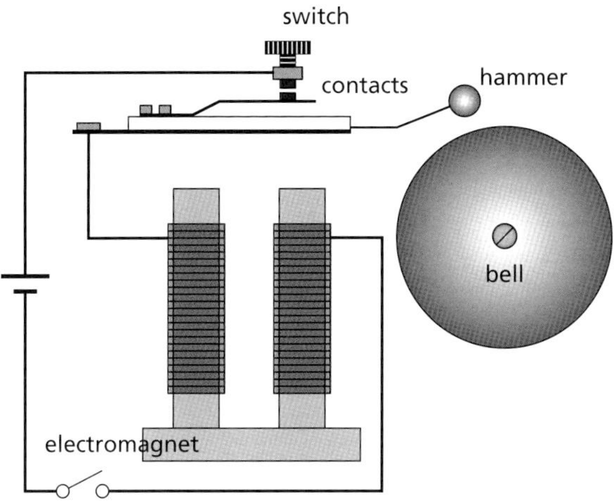

# End of stage review

1. Sarah walks to school. On the way she stops to talk to a friend. The graph shows her journey.

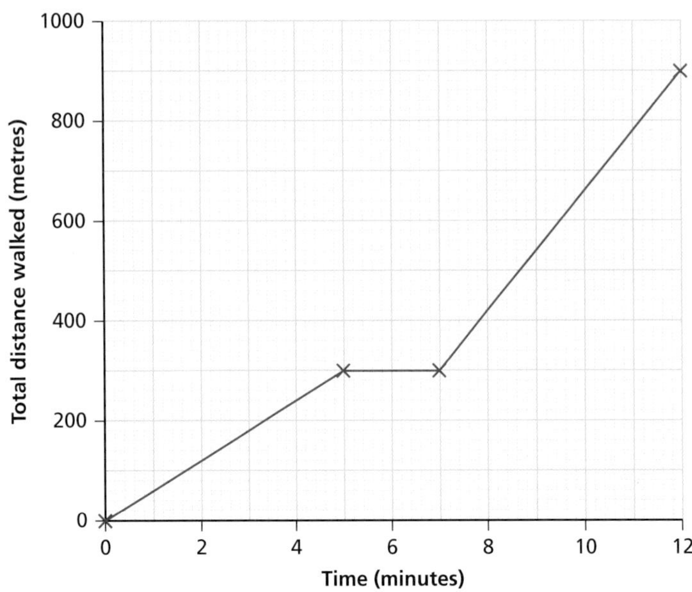


**(a)** Copy the table below. Use the graph to complete the table.

| | Time | Distance walked |
|---|---|---|
| First part of journey | ________ minutes | 300 metres |
| Talks to a friend | ________ minutes | 0 metres |
| Final part of journey | 5 minutes | ________ metres |

**(b)** Calculate Sarah's average speed in metres/second from home to school. Show how you worked out your answer.

2. Kayin sits on a see-saw. She weighs 500 N and sits 3 m from the pivot.

**(a)** Calculate Kayin's moment.

**(b)** Hani sits on the other side of the see-saw. He weighs 750 N and sits 2 m from the pivot on the opposite side. Calculate Hani's moment.

**(c)** Are Hani and Kayin balanced?

3. A box has a weight of 90 N and the side that is on the floor has a length of 1.2 m and a width of 0.5 m. Calculate the pressure that the box exerts on the ground.

4. The diagram shows what happens to white light when it enters and leaves a prism.

Complete the sentences below using *two* of these words.

| diffraction | reflection | away from | refraction | dispersion | towards |
|---|---|---|---|---|---|

When light enters a prism it changes direction. This is called ________________.

The light bends ________________ the normal when it enters the prism.

5. **(a)** What does 'primary colour' mean?

   **(b)** Name the three primary colours of light.

   **(c)** What would you see if you looked at a yellow T-shirt under green light?

6. **(a)** Copy and complete these sentences. Choose one of these words to complete each sentence.

| attract | repel |
|---|---|

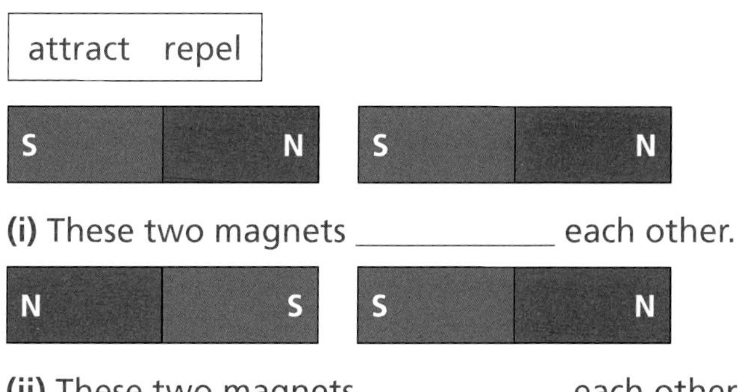


**(i)** These two magnets ____________ each other.

**(ii)** These two magnets ____________ each other.

**(b)** Write the letter of the correct drawing of the field lines around a bar magnet.

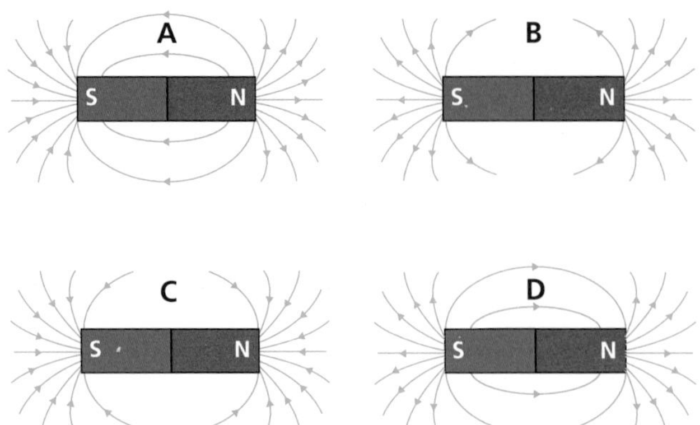


**(c)** Explain how Abhi could use an electromagnet to separate these aluminium and steel cans into separate containers.

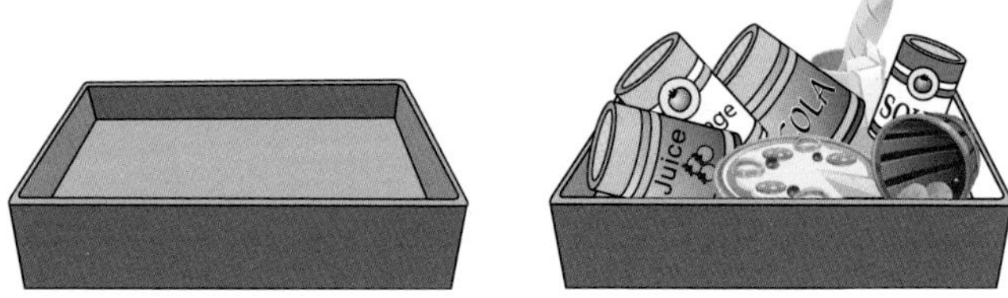

# Earth and space

*Chapter 11: The Earth and its resources*

11.1: Climate and weather 223
11.2: Climate change 230
11.3: Renewable and non-renewable resources 235
End of chapter review 244

*Chapter 12: Earth in space*

12.1: Asteroids 248
12.2: Stars and galaxies 253
End of chapter review 261
End of stage review 264

11

## Chapter 11

# The Earth and its resources

## What's it all about?

Earth is about 4.5 billion years old, and over that time the atmosphere has changed dramatically, life has evolved and the planet has become rich with different habitats for organisms and different resources. Humans like us have been present for just 200 000 years. During the past 200 years, the Industrial Revolution and a rapidly growing population have meant we are using many of Earth's resources so quickly that we may run out of them in the next 200 years. Meanwhile, we have learned that using fossil fuels to provide energy and materials is changing our climate, and this in turn may make Earth less able to sustain life. How to address these problems is a key challenge for us, and science is leading the way in finding answers.

You will learn about:

- The differences between climate and weather
- How the Earth's climate has changed naturally over long time periods
- How the Earth's climate has recently started to change much more quickly due to human activity
- How humans use the Earth's resources, and how we are learning to use new resources that can be replaced and affect the climate much less

You will build your skills in:

- Describing how a hypothesis can be proved or disproved by evidence
- Describing and interpreting trends and patterns in results
- Making predictions based on scientific knowledge and understanding

Chapter 11 . Topic 1

# Climate and weather

You will learn:

- The difference between climate and weather
- To understand the evidence for the natural cycle of Earth's climate
- To understand how long it takes for this cycle to take place
- To describe results in terms of any trends and patterns and identify any abnormal results
- To present and interpret scientific enquiries correctly
- To discuss how scientific knowledge is developed over time
- To describe the application of science in society, industry and research
- To use scientific understanding to evaluate issues
- To describe how scientific progress is made through individuals and collaboration
- To discuss the global environmental impact of science

## Starting point

| You should know that... | You should be able to... |
|---|---|
| The water on Earth goes through a cycle of evaporation, condensation, precipitation and return to the oceans | Use a range of secondary information sources to research and select relevant evidence to answer questions |
| The Earth is surrounded by its atmosphere, a mixture of gases including nitrogen, oxygen and carbon dioxide | Describe patterns in results |
| | Make a conclusion from results |

## The difference between climate and weather

You may have seen or heard weather forecasts and know about the topic of 'climate change'. You may even have heard some people say things like 'global warming can't be happening, because we still get really cold winters with lots of snow'. The problem with statements like that, is that they confuse two different things: 'weather' and 'climate'.

- **Weather** is what happens in the air around a local area for the next few hours or days.
- **Climate** is what happens in the air over very large areas (think of whole countries, continents or even the whole Earth) over long periods of time – years, hundreds of years, even thousands or millions of years.

It can help to understand the difference if we think about how to describe these things.

### Key terms

**climate**: what happens in the air over very large areas for long periods of time (many years).

**weather**: what happens in the air around a local area for a short time (hours or days).

When people say 'it's raining', 'it's very hot today', 'we have had a windy week' or 'it's going to snow next Tuesday', they are describing the weather.

When people say 'Our data for the last 200 years show that the average yearly temperature on Earth is increasing' or 'I would like to live in a country that is hotter and drier than this one', they are describing climate.

**11.1** *It is easy to observe how living things and their habitats are affected by the weather, but it is less easy to observe how they are affected by climate changes, because it takes many years to detect trends.*

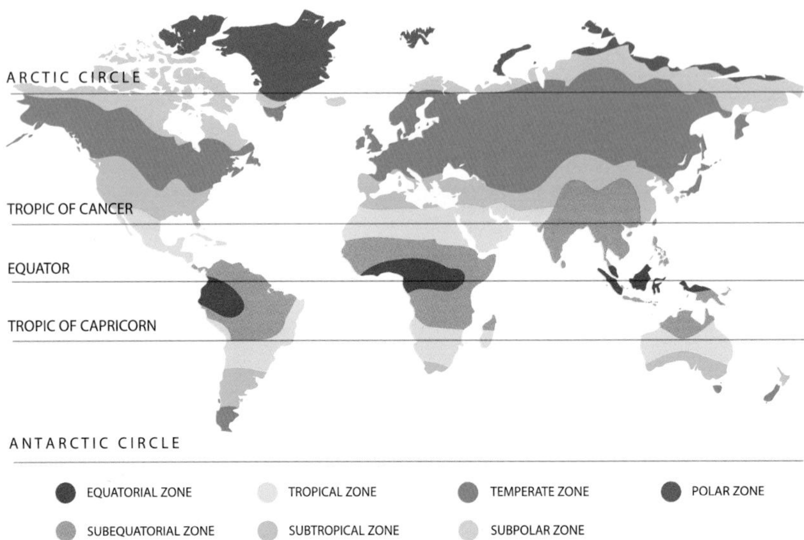


**11.2** *When we discuss climate, we are talking about average temperatures and weather conditions over large areas and periods of time. This map shows the different climate zones of Earth. There are hot dry areas, hot wet areas, cold dry areas and cold icy areas, and areas with climates in-between.*

## Activity 11.1: Describing weather and climate

Figure 11.3 is a photograph taken from a weather satellite. It shows part of North America (at the top of the picture), Central America (in the middle) and part of South America (at the bottom right). The huge swirl of cloud is a hurricane, which has thick clouds, heavy rain and strong winds.

**11.3** *Photograph taken from a weather satellite.*

**A1** Describe the weather affecting Central America in this photograph.

**A2** Suggest what kind of weather is affecting the part of South America that you can see in the photograph.

**A3** What can you say about the climate of Central America? For example, is it correct to say that the climate is 'hurricanes'?

**A4** Challenge Describe how measurements of the weather in Central America can be used to build up an idea of the climate there.

1 Complete the sentences by using the words 'weather' and 'climate'.

A scientist measuring the temperature outside their laboratory can use the results to describe the __________ .

A team of scientists from different places recording the temperature every day for 20 years and calculating the average yearly temperature can use the results to describe the __________ .

2 Look at Figure 11.4. It shows a graph of the amount of rain that has been collected during October, in one place, every year for 20 years.

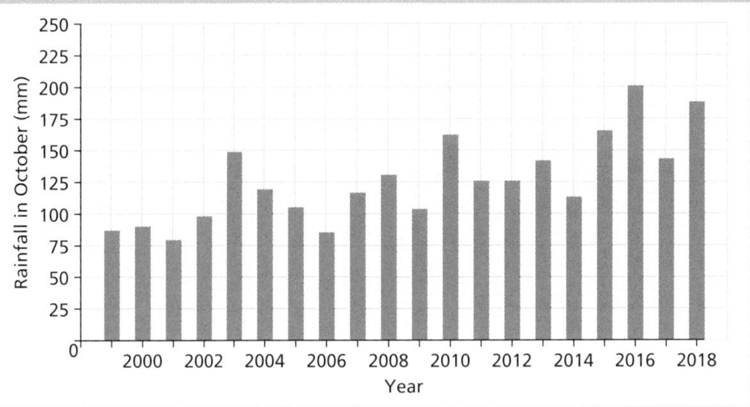


**11.4**

**a)** How much rainfall was measured in October 2016?

**b)** What does this tell you about the weather in October 2016 compared to October 2012?

**c)** Is there a trend in the measurements? If so, describe it.

**d)** What extra information would you need to describe the climate of the country where these measurements were taken?

3 Someone tells you that the climate cannot be getting warmer because they just experienced the coldest day in March that they can remember. Explain why their example is not good evidence for or against climate change.

## Natural changes in Earth's climate

You should remember that the mixture of gases in the Earth's atmosphere has changed naturally over many millions of years. These changes in gases affect the climate. There are also other natural cycles in Earth's climate, caused by a number of other changes which include:

- changes in the energy produced by the Sun
- increases and decreases in the number of active volcanoes
- small changes in the Earth's orbit around the Sun.

The most significant of these cycles is called the **ice age cycle**. Scientists think this is mainly related to small changes in the Earth's orbit around the Sun.

During an **ice age**, these small changes in the Earth's orbit cause the energy from the Sun to reach different parts of the Earth's surface. Some areas receive more energy and others receive less.

There have been at least five major ice ages in Earth's history (the Earth is about 4.5 billion years old). Within each major ice age, there are periods of time where the Earth is colder and ice covers large parts of the Earth. These are called **glacial periods**.

There are also warmer periods where less ice covers the Earth. These are called **interglacial periods**. Glacial periods take place in cycles of 100 000 years and 40 000 years, so understanding how ice ages work and when they have happened is quite complicated.

Scientists have a number of ideas about what causes glacial periods and interglacial periods within an ice age, but there is no single, simple reason.

The overall effects as a glacial period begins are that:

- the **ice sheets** at the North and South Poles spread much further, as the land and sea around the poles get colder
- **glaciers** in mountain areas thicken and get longer
- large amounts of carbon dioxide in the atmosphere become 'trapped' inside the ice
- this causes the atmosphere to cool, so that the spread of ice increases
- the average temperature of the Earth's atmosphere near the surface reduces by 5–8 °C.

This cooling takes around 10 000 years. The glacial period can last for 20 000 years or up to 80 000 years, depending on the cycle involved. Then, as the Earth's orbit changes back to

### Key terms

**glacial period**: time during an ice age when ice covers more of the Earth and the climate is cooler; it can last between 20 000 and 80 000 years.

**glacier**: slow-moving ice 'river' formed from snow squeezed together over more than one hundred years.

**ice age**: period of time when the Earth is several degrees Celsius colder than usual, due to changes in the Earth's orbit.

**ice age cycle**: cycle of Earth's climate changing from warm period to ice age and back again.

**ice sheet**: large covering of ice at the North and South Poles of Earth.

**interglacial period**: time during an ice age when ice covers less of the Earth and the climate is warmer.

its more normal state, the Sun's energy warms the Earth more, the ice melts, carbon dioxide is released back into the atmosphere and the Earth's temperature returns to where it was before the glacial period. This warming takes around 10 000 years.

The end of the last glacial period on Earth was around 11 700 years ago. Over the following 10 000 years, temperatures returned to those we are more familiar with today.

Earth is currently in the interglacial period of the Quaternary glaciation ice age.

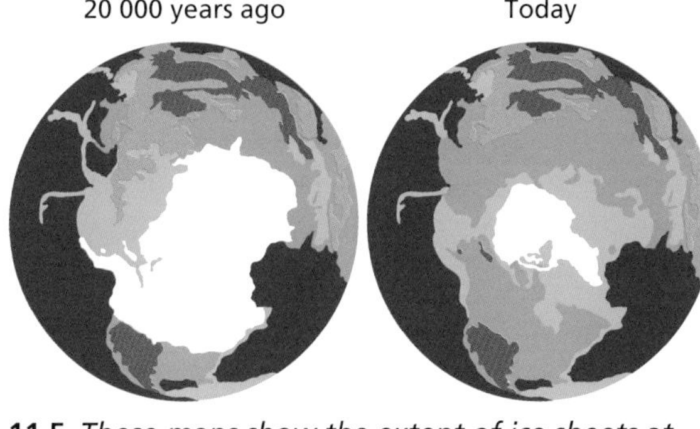


**11.5** *These maps show the extent of ice sheets at the North Pole 20 000 years ago in the last glacial period, compared to today.*

Scientists have found many kinds of evidence that supports the idea of regular ice ages and glacial periods and how often they occur.

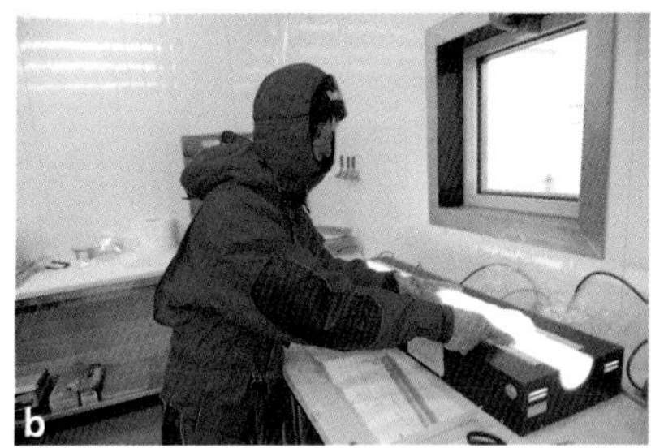

**11.6** *Evidence for glacial periods on Earth and when they occurred comes from:* **a** *valleys caused by the movement of glaciers,* **b** *ice core samples and* **c** *animal fossils and skeletons.*

- Effects of ice on rock: the growth of ice sheets and glaciers causes mountains to be worn away to form valleys, and rocks to be carried by the ice to places they would not normally be found. These valleys and rocks are present in places very far away from today's polar ice and glaciers.
- Ice cores: scientists take samples called **ice cores** from deep in the ice and analyse the frozen bubbles of gas and other substances found in the ice. Because the ice formed in glaciers and ice sheets builds up over thousands of years, it is possible to find out the conditions from long ago. For example, ice cores taken from many different places around the world tell us that the Earth was colder 20 000 years ago than it is today.
- Animal fossils and skeletons: during glacial periods, animals tend to move away from the polar regions towards the Equator, where it is warmer. Bodies of dead animals that are buried under sediments turn into fossils. The places where we find these fossils tell us where some species of animals moved during the last glacial period. We have also found preserved skeletons of animals that are not found on Earth today, because they lived in much colder temperatures.

### Key term

**ice core:** sample taken from deep in the ice and analysed scientifically.

## Science in context: Finding evidence from ice

Scientists write hypotheses to try to explain why the things we observe around us happen in the way that they do. The most important part of the scientific process then takes place: the search for evidence, which can be found either to agree with or to contradict the hypothesis. When enough evidence is collected, the hypothesis can be improved or replaced with a better hypothesis.

In looking for evidence of the past history of Earth, scientists discovered that ice can tell us a lot about things that happened on Earth many thousands or tens of thousands of years ago.

Figure 11.6b shows a scientist taking measurements of an ice core sample. Ice sheets and glaciers build up over many thousands of years, from snow falling and settling in very cold locations. Over hundreds of years, the layers of snow are squashed together by the weight of all the newer snow above. This forms ice layers that are buried so deeply that the air above the glacier or ice sheet does not reach them.

Bubbles of air trapped in the ice as it forms contain samples of the Earth's atmosphere from thousands of years ago. Scientists today drill deeply into ice sheets and glaciers. The bottom of the ice core they remove contains air that can be analysed to show what the Earth's atmosphere was like at that time. All ice cores are stored so they stay frozen, so that scientists from across the world can take small samples and gather more evidence about the past.

### Making links

Ecosystems on Earth contain a wide variety of habitats for plants and animals (Chapter 3).

**a** How does climate affect habitats and the organisms within them?

**b** Suggest how ecosystems are affected during glacial periods, and what may happen to organisms as a result.

## Activity 11.2: Investigating the Woolly Mammoth

Figure 11.6c shows the skeleton of an animal called a Woolly Mammoth. It lived on Earth until about 10 000 years ago.

**A1** Find a picture of what scientists think a Woolly Mammoth looked like. Make your own drawing and label three features that helped it to live in the cold.

**A2** The Woolly Mammoth is now extinct. Explain what 'extinct' means.

**A3** Describe how the Earth's temperature started to change about 11 700 years ago in a way that affected the Woolly Mammoth.

**A4** Describe the difference between a glacial period and an interglacial period.

**A5** Challenge Do you think Woolly Mammoths could live on Earth today? Explain your thinking.

4. Name the *three* types of evidence that show how the Earth experiences glacial periods during the ice age cycle.

5. Describe a glacial period. What happens during the glacial period? How long does one take?

6. Predict when the next glacial period might occur.

7. Scientists have found lots of evidence that the average temperature of the Earth's surface has been increasing over the past 100–200 years. It has changed by 0.9–1.0 °C.

   **a)** Compare this to the change in temperature since the last glacial period.

   **b)** Compare the timescale of this change over the past 100–200 years to the timescale of glacial periods. Do you think the recent change is due to a glacial period? Explain your thinking.

## Key facts:

- Weather describes the conditions in the atmosphere over a small area at a particular time.
- Climate describes the conditions in the atmosphere over a large area, averaged over a long time.
- Earth's climate changes naturally over time, involving warm periods and ice ages.
- The natural cycle of changes in Earth's climate takes place over very long time periods, typically tens of thousands of years (for glacial and interglacial periods) and hundreds of millions of years (for ice ages).

## Check your skills progress:

- I can describe the evidence for the natural changes in Earth's climate, including effects on the landscape, ice cores and fossils.
- I can identify trends in measurements of the Earth's climate.
- I can interpret observations and measurements of the Earth's climate.

# Climate change

You will learn:

- To understand what can cause changes in the Earth's climate
- To identify a hypothesis as testable
- To describe how evidence affects scientific hypotheses
- To plan investigations, including fair tests, while considering variables

## Starting point

| You should know that... | You should be able to... |
|---|---|
| Earth's atmosphere contains approximately 78% nitrogen, 21% oxygen and 0.03% carbon dioxide | Use a range of secondary information sources to research and select relevant evidence to answer questions |
| The mixture of gases in the Earth's atmosphere has changed very gradually, over hundreds of millions of years | Describe patterns in results |
| The Earth's climate goes through a cycle, from warm periods to ice ages and back again, over many thousands of years | Present and interpret results using bar charts and line graphs |

## The effect of gases in the atmosphere on climate

You should remember that Earth's atmosphere is a mixture of gases, including nitrogen, oxygen and carbon dioxide. Even though it makes up only about 0.03% of the atmosphere, carbon dioxide is a very important gas. It can absorb energy and reflect it back.

Life on Earth depends on energy provided by the light of the Sun. When it reaches the ground, this energy heats up the ground and the water in the oceans. The ground and water reflect or re-emit most of this energy into the atmosphere. The carbon dioxide in the air 'traps' and reflects this energy back to the ground, so that over time, the surface of Earth is kept warm (figure 11.7).

We call carbon dioxide a **greenhouse gas**, because it acts like the glass panels in a greenhouse that keep plants warm and help them grow. Methane (natural gas), sulfur dioxide

### Key term

**greenhouse gas**: gas in the atmosphere that reflects energy back to the surface, causing the temperature to rise.

and nitrogen oxides are also greenhouse gases, but they are present in much smaller quantities than carbon dioxide.

Without carbon dioxide in the atmosphere, Earth's surface would be too cold for most of the plants and animals that live there.

However, if the amount of carbon dioxide in the atmosphere changes, then the amount of energy reflected back to the surface changes too. This, in turn, changes the average temperature of the atmosphere and the surface, which causes Earth's climate to change.

You have seen how, during a glacial period, carbon dioxide is 'trapped' in large areas of ice, meaning that less carbon dioxide is in the atmosphere, and the Earth cools further as a result. This natural change in temperature of 5–8 °C takes around 10 000 years. However, if the Earth warms and some of the ice melts, then the trapped carbon dioxide is released into the atmosphere, increasing the amount of atmospheric carbon dioxide and helping to speed up further warming.

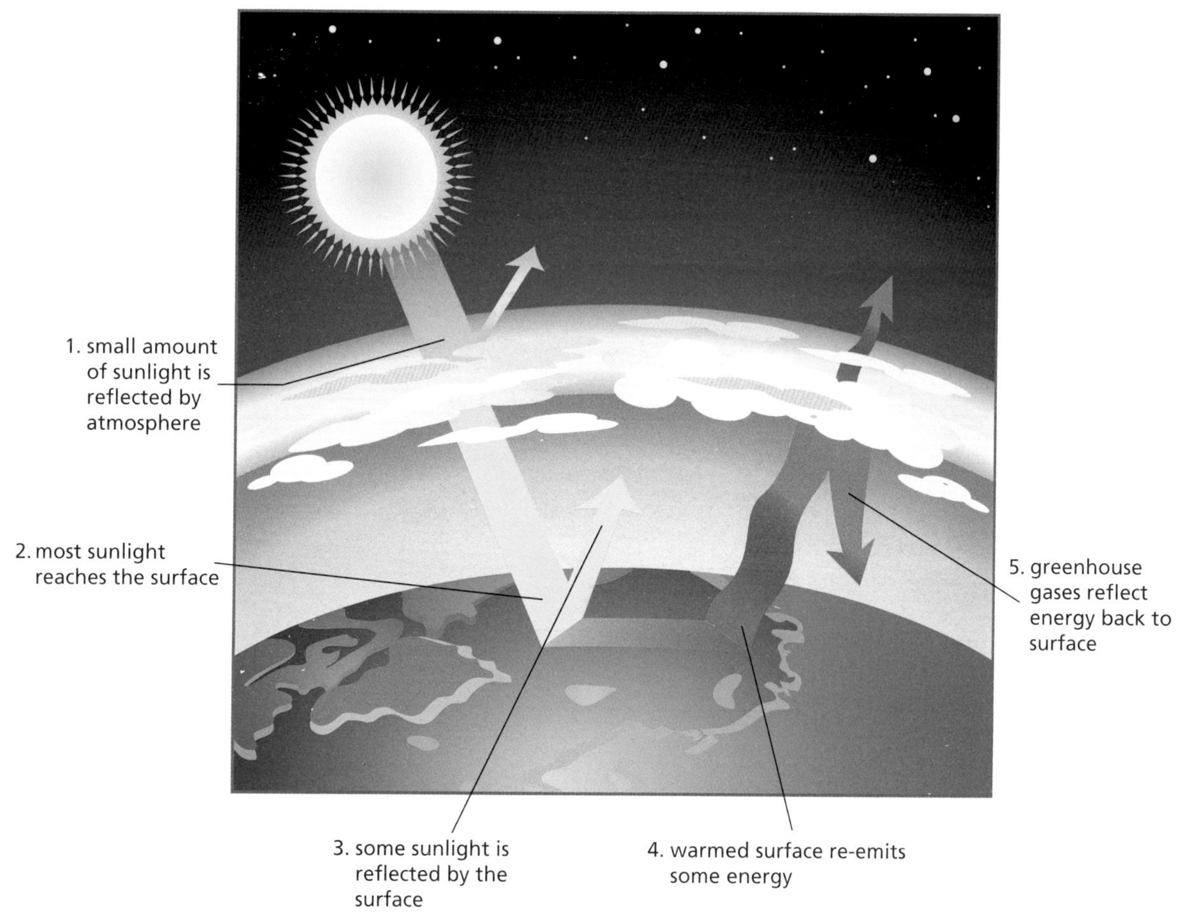


**11.7** *Greenhouse gases such as carbon dioxide reflect energy back to the Earth's surface, keeping the Earth warm.*

### Making links

The mixture of gases in the atmosphere can be tested (Stage 7, Chapter 10).

Describe how you could determine the amount of oxygen in a sample of air.

1. Which of the four gases present in significant amounts in the Earth's atmosphere is a greenhouse gas?
   **a)** nitrogen
   **b)** oxygen
   **c)** argon
   **d)** carbon dioxide
2. Describe what a greenhouse gas does in the Earth's atmosphere.
3. If the amount of greenhouse gases in the atmosphere increases, what will happen to the Earth's climate?
4. Would life exist on Earth if there were no greenhouse gases in the atmosphere? Explain your answer.

## Human activities that affect the atmosphere

Many processes that humans have developed release gases into the atmosphere. The easiest of these to understand is combustion. **Combustion** describes the chemical reaction that takes place when a substance is burned so that it reacts with oxygen. We often use this reaction to provide energy. When the substance used in combustion is a **fuel**, the reaction releases a lot of energy, which we can then use to power processes and devices. Examples include:

- burning coal or gas in power stations, to heat water that turns generators to make electricity
- burning gas or oil in our homes to heat rooms or to cook food
- burning petrol or diesel in engines to make vehicles such as cars, trucks and buses move.

### Key terms

**combustion**: reaction of a substance with oxygen.

**fuel**: substance that releases energy.

**11.8** *Eggborough power station, Yorkshire, England.*

In these examples, all the fuels mostly contain carbon and hydrogen. They are called **hydrocarbons**. The general word equation for their combustion is

hydrocarbon fuel + oxygen (from the air) → carbon dioxide + water + [energy]

These hydrocarbon fuels are also referred to as fossil fuels (see topic 11.3). As you can see, these reactions produce carbon dioxide, which goes into the atmosphere. In small amounts, this does not affect the total carbon dioxide in the atmosphere very much. However, now that these reactions are being used to produce energy all over the Earth, the total amount of carbon dioxide produced is large enough to change the mixture of gases in the atmosphere.

Every year for the past 100–200 years humans have been burning more and more fuels, so the amount of greenhouse gases in our atmosphere is increasing, leading to an increase in the Earth's average temperature. These changes are often referred to as **climate change**.

This increase is happening over a much smaller timescale than the natural cycles of Earth's climate. Scientists are worried about this, because the models they use to predict the future climate all agree that the Earth's temperature is going to increase by 1.5–4 °C in fewer than 100 years. We don't know exactly what all the effects will be, but we do know that:

- ice at the Earth's poles is already melting faster than usual
- sea levels are already rising due to melting ice and will continue to rise
- there will be more extreme weather events such as hurricanes, droughts and flooding.

Scientists call this process **global warming**. Teams of scientists and engineers across many countries are working to develop ways of reducing the amount of greenhouse gases we put into the atmosphere. This should help us to limit the amount of temperature change.

a

b

**11.9** *Climate change is linked to* ***a*** *an increase in flooding in some areas and* ***b*** *an increase in drought in other areas.*

### Key terms

**climate change**: changes to the average long-term temperature and weather patterns of the Earth.

**global warming**: increasing temperatures on the Earth and in its atmosphere.

**hydrocarbon:** substance made only of carbon and hydrogen.

### Activity 11.3: The changing work of climate scientists

Climate scientists work in large teams, often spread across the world, sharing ideas and results using the internet. They study data from satellites orbiting the Earth. The satellites collect many different types of data including land and sea temperatures, areas of sea ice and weather patterns.

The main hypothesis of recent climate science is that the average temperature of Earth's surface is increasing more rapidly than is natural, due to human activities.

**A1** Is the main hypothesis of recent climate change testable? Give reasons for your answer.

**A2** Write a short paragraph outlining an investigation your class could undertake that would gather some evidence to test this hypothesis.

**A3** Explain why climate scientists need to take measurements in many different places and over long periods of time.

**A4** Research a scientist or group of scientists who are studying climate change.

**a)** What are the main aims of this scientist or group?

**b)** What are their main findings to date?

**c)** What could we do to support the work of these scientists?

**5** Explain why some fuels are called 'hydrocarbons'.

**6** Name *three* fuels containing mostly hydrocarbons.

**7** Name a gas that causes global warming.

**8** Write the word equation for the combustion of a hydrocarbon.

**9** Explain why some scientists think we should stop burning hydrocarbons.

**10** Even if we stopped all use of hydrocarbon fuels this year, the average temperature of Earth's surface will still rise for the next few years. Suggest why this would happen.

## Key facts:

- The mixture of gases in the Earth's atmosphere can change over time.
- Changes in the amounts of some gases in the Earth's atmosphere, particularly carbon dioxide, cause changes in the Earth's climate.
- The Earth's climate is changing more rapidly than is natural because of human activities.
- These changes include rising average temperatures, melting polar ice, rising sea levels and more extreme weather events.

## Check your skills progress:

- I can identify whether the main hypothesis about recent climate change is testable.
- I can plan an investigation to help gather evidence about climate change.

Chapter 11 . Topic 3

# Renewable and non-renewable resources

You will learn:

- To identify renewable and non-renewable resources
- To describe how humans use renewable and non-renewable resources
- To sort and group materials using secondary information
- To make predictions based on scientific knowledge and understanding
- To use results to describe the accuracy of predictions
- To describe results in terms of any trends and patterns and identify any abnormal results
- To reach conclusions studying results and explain their limitations
- To discuss how scientific knowledge is developed over time
- To describe the application of science in society, industry and research
- To use scientific understanding to evaluate issues
- To describe how scientific progress is made through individuals and collaboration
- To discuss the global environmental impact of science

## Starting point

| You should know that... | You should be able to... |
|---|---|
| Earth is the source of all the materials that humans use and many of these materials come from or are found in rocks | Use a range of secondary information sources to research and select relevant evidence to answer questions |
| Fossils can form in sedimentary rocks | Describe patterns in results |
| Word equations can be used to describe chemical reactions | Make a conclusion from results |

## Resources

Over time, humans have developed many different ways of making new things from the materials and processes that occur naturally on Earth.

One of the biggest advances for humans was when we discovered how to produce electricity, which we now use to power all kinds of devices.

Another advance was the development of the internal combustion engine, which burns fuel to make vehicles move. Skyscrapers, houses, roads and schools are evidence of our use of materials to build things. More recently, rarer materials have been used to make parts of electronic devices.

All these things that humans make require **resources**. Fuels and rock are taken from the ground. Lots of different chemical reactions are used to take metals out of rocks. Energy from the Sun and wind is collected to make electricity.

Some of these resources are **renewable resources**. These resources are easily replaced over short timescales (a few years at most). Examples of renewable resources include:

- crops such as wheat, soybean and sugarcane, which can be turned into fuels but replaced each year
- energy from the Sun, including light used by solar panels to produce electricity
- water, where the energy from moving water can be used to produce electricity, using hydroelectric generators or tidal generators
- wind, where the movement of the air turns devices that produce electricity
- geothermal energy, which uses energy from hot rocks deep underground.

**11.10** *Sugarcane is grown for food but also for use as a fuel.*

There are also many **non-renewable resources**. These resources either cannot be replaced or can only be replaced over very long timescales, often millions of years. These resources are finite, meaning they will eventually run out. Examples of non-renewable resources include:

- rock and stone taken from the ground to build houses, office blocks, bridges and roads
- hydrocarbon fuels such as coal, gas and oil, and petrol and diesel, which are made from oil
- fuel for nuclear power stations, although only small amounts of fuel are needed to produce large amounts of energy
- rare metals such as gallium, indium and yttrium, which are used to make parts of devices such as mobile phones.

**11.11** *Different types of rock are dug out of the ground for the construction of buildings, bridges and roads.*

**1** Describe the difference between a renewable resource and a non-renewable resource.

**2** The Sun uses hydrogen as a nuclear fuel. Suggest why sunlight is considered to be a renewable resource.

### Key terms

**non-renewable resource**: resource that cannot be replaced easily and will run out one day.

**renewable resource:** resource that can be replaced easily in a short time.

## Non-renewable resources: fossil fuels

**Fossil fuels** are non-renewable resources that are formed from the remains of dead animals and plants.

Sometimes, when plants died in swampy areas, they would be covered with mud and sand. When this happened, the remains were not surrounded by oxygen-containing air, so the plants did not decay quickly. Instead the remains became buried under more material. As the remains were squeezed due to the weight of everything above, heat energy coming from the Earth converted them to coal.

This also happened under the sea. When sea creatures died they sank to the bottom of the sea where they were buried under sand. The pressure and heat turned the remains into oil.

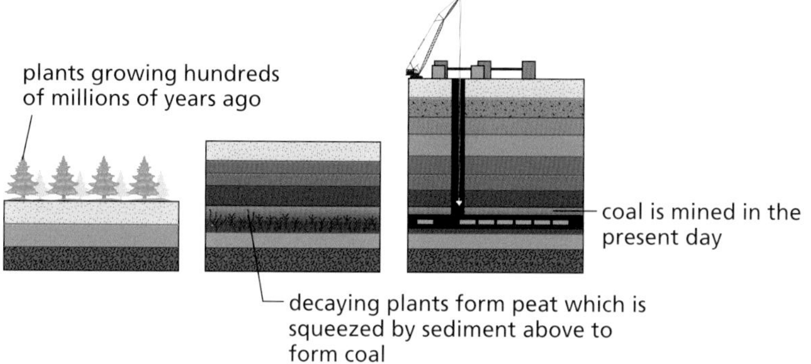


**11.12** *Coal is formed over hundreds of millions of years by the squeezing and heating of dead plants.*

Fossil fuels are relatively cheap to extract and transport. They are also efficient – they release a lot of energy when you burn them. They do not rely on climate or location, unlike solar energy and geothermal energy, respectively.

Burning fossil fuels produces a lot of pollutants. One of these is carbon dioxide, a greenhouse gas which leads to climate change, as described earlier. Other pollutants are also produced.

- Burning fossil fuels releases waste gases including sulfur dioxide and nitrogen oxides. In the air these gases dissolve in water droplets forming acid rain. Acid rain harms sea life and wildlife on land, and damages land and buildings.
- Burning coal and diesel produce some solid substances, such as soot (carbon), in the form of very small particles that are carried up into the atmosphere. These particles collect on buildings, causing them to gradually turn darker in colour. The particles can also be breathed in

### Key term

**fossil fuels**: non-renewable resources that are formed from the remains of dead animals and plants.

**11.13** *Acid rain killed these trees in Germany.*

by people, causing illnesses of the respiratory system such as asthma and bronchitis.

Because burning fossil fuels has many disadvantages, we are increasing our use of other energy sources.

### Activity 11.4: Investigating pollutant particles

Does the burning of hydrocarbon fuels cause pollution in your area?

Hang a white tissue or piece of fabric outside for a week in a position that is sheltered from rain. Then compare it with one that has been kept inside.

**A1** Make a prediction before you start the investigation and explain why you think this.

**A2** Say whether your results match your prediction.

**A3** Suggest reasons why your prediction was accurate or not.

**A4** Use your results as evidence to make a conclusion.

**A5** What piece of equipment could you use to examine particles more closely?

**A6** Present your findings as a short report. Do not use more than 200 words.

**3** Describe how fossil fuels are formed.

**4** Explain why gas is a non-renewable resource.

**5** Describe an advantage of using fossil fuels to produce energy.

**6** Give some disadvantages of burning coal.

**7** Coal is still the world's largest source of energy for producing electricity despite its disadvantages. Why is this?

## Non-renewable resources for purposes other than energy

We also use many non-renewable resources for other purposes.

- We remove many materials from the ground such as limestone and gravel to construct buildings, bridges and roads.
- We extract metals from rocks dug out of the ground, for buildings and to construct devices.

We take these resources continuously and they are never replaced. The removal of these resources, and the machinery and transportation used to move the resources to where they are needed, can cause significant damage to habitats of plants and animals.

**Key term**

**plastic materials**: materials derived (made) from oil and processed so they can be moulded.

**Plastic materials** are used in all sorts of different ways:

- for packaging
- for making parts of devices
- for bottles and boxes
- for decorative materials
- for bags and some clothing
- for toys
- for furniture
- for pens and pencils
- as tiny beads in some cosmetic products, such as skin creams.

Plastics are produced mainly from oil. Oil is a mixture of different hydrocarbons, and some of these hydrocarbons are separated out and used in reactions to produce plastic materials. An example is poly(ethene) – you may know it as 'polythene' – which is used to produce plastic bags.

**11.14** *We throw away large amounts of plastic waste every day, and most plastics do not decompose.*

Plastic materials are very useful, but cause a problem when they are thrown away after use. Plastics made from oil cannot be broken down by decomposers such as bacteria. If they are collected in household waste and stored, they can last up to 1000 years and are harmful to animals that may eat them. If they are thrown away or tipped into water, they end up in the Earth's rivers and oceans, trapping animals and harming those that ingest (swallow) them.

Plastic beads in some cosmetic products can also get into the water system, for example, when someone rinses off the product and the water containing plastic beads drains away. Eventually this water will reach the ocean, and the plastic beads build up in the ocean. Animals such as sea turtles may swallow the beads, and they cannot get rid of them. Eventually animals that swallow enough beads will die.

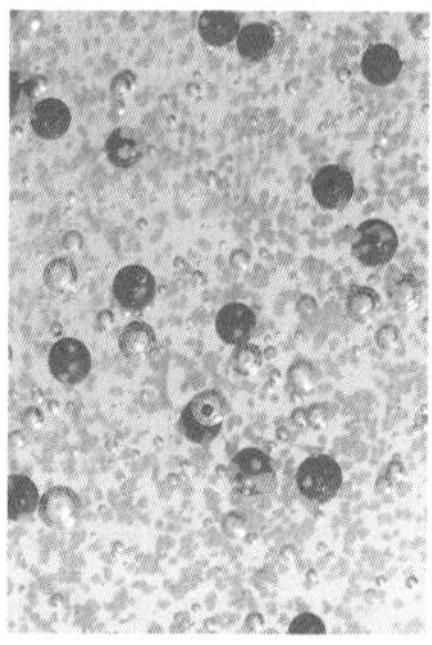

**11.15** *Tiny plastic beads are used in many cosmetic products, but after use they collect in the ocean where they can harm animals.*

## Activity 11.5: Investigating the use of plastics in the classroom

How many plastic items are used in your classroom?

Form groups of three or four. When the signal is given, you have three minutes to identify and collect or describe as many items made of plastic as you can find in the classroom.

**A1** Make a prediction before you start the investigation of how many items you think you will find.

**A2** Say whether your results match your prediction.

**A3** Suggest reasons why your prediction was accurate or not.

**A4** Use your results as evidence to make a conclusion.

**A5** Were there any items made from bioplastics? (Bioplastics are used in the same way as plastics made from oil but they are derived from renewable sources and are able to decompose.)

**A6** Present your findings as a short report. Do not use more than 200 words.

**8** Give *three* ways in which humans use non-renewable natural resources.

**9** Give *four* uses of plastic materials.

**10** Describe *two* ways in which plastic waste can affect animals.

**11** Explain *three* ways in which a growth in the human population increases the damage to habitats caused by the use of non-renewable resources.

## Renewable resources for energy

Governments around the world have agreed that we should all reduce our use of fossil fuels. Many are looking at alternative ways of making electricity, heating buildings and running vehicles.

We now use some plants and algae to make **biofuels** for vehicles. But this means farmers use land for growing biofuel plants instead of growing food for humans. When burned, biofuels release some carbon dioxide into the air, but not as much as fossil fuels do.

Ways of producing electricity using renewable resources are becoming more common. They include power from:

- wind
- waves
- tides
- the Sun (**solar power**)
- geothermal heat sources
- water flowing out of dams (**hydroelectric power**).

These do not release gases into the air and so cause less pollution.

**11.16** *Some aircraft now burn biofuels.*

### Key terms

**biofuels**: fuels made from renewable sources such as crops.

**hydroelectric power**: electricity produced using energy from moving water as it is released from behind a dam.

**solar power**: electricity produced using energy from sunlight.

## Activity 11.6: The world's energy consumption

We call the total amount of energy used by humans in a year 'the world's energy consumption'. This increased rapidly in the 20th century. However, the sources of energy we are using have changed. Figure 11.17 shows the world's energy consumption in 1000s of terawatt-hours (TWh) per year. 1 TWh = 3 600 000 000 000 000 J. In standard form this is $3.6 \times 10^{15}$ J.

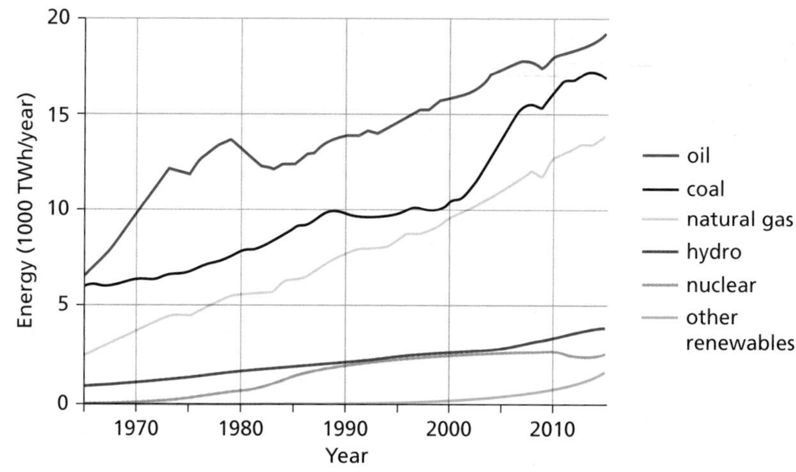


**11.17** *The world's energy consumption since 1965.*

During the 20th century, most of our electricity was generated using non-renewable energy sources. Due to the increase in energy usage, we have increased the use of renewable energy sources in the generation of electricity.

**A1** List the sources of energy that must be included in 'other renewables'.

**A2** Look at figure 11.17. What is the trend of *all* the energy sources from 1965 to 2015?

**A3** Suggest a reason for this trend.

**A4** Look at figure 11.17. Which energy source has the world used *less* of since 2000?

**A5** Use your knowledge of renewable and non-renewable energy sources to predict which energy sources are likely to:

**a)** increase in the future

**b)** decrease in the future.

**A6** Challenge Explain *one* of your predictions.

## Science in context: The cost of renewable resources

Developing new technology is often expensive. For example, the first 'solar cells' used to store energy from sunlight and provide electricity were made in 1883. They transferred less than 1% of the energy they collected as useful energy. They were made using the expensive elements selenium and gold.

As more people became interested in the new technology, different ways were invented of making these solar cells using different, cheaper resources. In the last 20 years, countries across the world have installed more solar cells, in order to help reduce the use of fossil fuels. In 2011, Chinese companies could produce solar cells that cost just $1.25 per watt of power produced. In 2018, a company based in the United States produced a solar cell that transferred 29.1% of the energy it collected as useful energy.

As the cost of technology reduces and the efficiency of the technology improves, countries and companies switch to use more and more energy resources that are renewable.

**12** Is wave power renewable or non-renewable? Give a reason for your answer.

**13** **a)** Give *one* advantage of solar power compared to burning fossil fuels.

**b)** Suggest *one* disadvantage of solar power.

**14** **a)** Describe *one* way in which wind power is a benefit for the environment.

**b)** Describe *one* way in which wind power may harm the environment.

**15** Explain how using satellite tags to track birds is useful for deciding where to build wind turbines.

**11.18** *Wind turbines are a renewable source of energy, but they only produce electricity if there is wind. Turbines spin round when the wind blows and turn a generator to produce electricity. They may also be dangerous for birds flying past in large flocks, they may damage habitats and some people think they are ugly.*

## Renewable resources for purposes other than energy

Wood is a renewable resource, but some trees grow quickly and more easily than others. This means that wood taken from cutting down these trees is more quickly and easily replaced than wood taken from trees that take longer to grow. As long as people or companies taking the wood plant more trees, and use only faster-growing trees, wood is an important renewable resource used in construction.

Straw and bamboo can be used as building materials. They are naturally strong and light, and are easily grown again to replace the materials used.

**11.19** *Wood is used increasingly often in modern buildings because it is a renewable resource.*

**Bioplastics** have been developed, which are plastic materials produced from renewable sources such as vegetable fats and oils, straw, wood and recycled food waste. Some bioplastics can be produced from waste objects made from certain types of plastic, such as soft drink bottles, by using bacteria to decompose the plastic quickly.

Key term

**bioplastic**: a material that can be used in the same way as plastics produced from oil, but which is produced using renewable sources.

**16** Suggest *three* renewable materials that can be used in construction.

**17** **a)** Give *one* advantage of wood as a construction material.

**b)** Suggest *one* disadvantage of wood as a construction material.

**18** Explain why bioplastics are made from renewable sources, but 'normal' plastics are from non-renewable sources.

**19** Suggest why bioplastics do not present a significant problem when they are thrown away.

**11.20** *This lunch box, fork and spoon are produced using bioplastic that is made from starch, the white powdery substance in the picture, which is found naturally in crops such as potatoes, wheat and rice.*

## Key facts:

- ✔ As humans increase their population and develop new technology, they use more and more resources for energy and for making things.
- ✔ Renewable resources are resources that can be replaced easily in a short space of time.
- ✔ Non-renewable resources cannot be replaced easily and will run out one day.
- ✔ The use of many non-renewable resources, such as fossil fuels, causes pollution.
- ✔ The combustion of fossil fuels produces carbon dioxide gas, which causes climate change.
- ✔ We need to find and develop more renewable resources and use these instead of non-renewable resources, to reduce pollution and other damaging effects of non-renewable resources.

## Check your skills progress:

- I can sort and group information.
- I can make and test predictions about investigations into air pollutants.
- I can describe trends and patterns in results.
- I can make conclusions by interpreting results.

# End of chapter review

## Quick questions

1. If I say it is raining, I am describing:
   - **(a)** weather
   - **(b)** climate
   - **(c)** both weather and climate
   - **(d)** neither weather nor climate.
2. A glacial period is a time when:
   - **(a)** the Earth warms and ice cover reduces
   - **(b)** the Earth warms and ice cover increases
   - **(c)** the Earth cools and ice cover reduces
   - **(d)** the Earth cools and ice cover increases.
3. Climate change caused by increased carbon dioxide due to human activities is thought to have taken place over the last:
   - **(a)** 10 years
   - **(b)** 100 years
   - **(c)** 1000 years
   - **(d)** 10 000 years.
4. Which of the following is a non-renewable resource?
   - **(a)** gas
   - **(b)** hydroelectricity
   - **(c)** biofuel
   - **(d)** wood
5. Bioplastics can be made from:
   - **(a)** vegetable oil
   - **(b)** starch from potatoes, rice or wheat
   - **(c)** recycled food waste
   - **(d)** all the above.

6. **(a)** Describe *two* differences between weather and climate.

   **(b)** A scientist says that 'the Earth's climate has been getting warmer for the last 100 years'. Describe what this means.

7. **(a)** Name and describe *one* type of evidence that the Earth has been through colder conditions tens of thousands of years ago.

   **(b)** Describe *two* changes that could be observed on Earth as a result of the temperature change during the last glacial period.

8. Coal is burned to produce electricity.

   **(a)** Name the greenhouse gas that is produced in large amounts during this process.

   **(b)** What is the main effect of increasing the amount of this gas in the atmosphere?

   **(c)** Name *two* other types of pollution that can occur as a result of burning coal to generate electricity.

## Connect your understanding

9. **(a)** Explain what 'renewable' means.

   **(b)** Which of the following resources used for construction are non-renewable?

   **rock** **wood** **bamboo** **metal** **straw**

   **(c)** Describe *two* problems with using non-renewable resources in construction.

10. Crude oil is a mixture of hydrocarbons taken from deep underground.

    The hydrocarbons in crude oil can be separated. Several of these hydrocarbons are fuels that can be burned to produce energy.

    **(a)** Name *three* hydrocarbon fuels and state where each is used.

    **(b)** Other products made from crude oil include plastics. Describe *two* disadvantages of using plastic to make products.

    **(c)** Suggest *one* alternative material that could be used in place of plastics, and describe what it is produced from.

11. A scientist investigating evidence for climate change examines an ice core.

    **(a)** Describe an ice core and where it is taken from.

    **(b)** Why does an ice core have to be stored at temperatures below 0 °C?

    **(c)** Suggest what an ice core can tell the scientist about the Earth's climate.

**12.** Biofuels are an important resource.

**(a)** Describe what a 'biofuel' is.

**(b)** Explain why biofuels are renewable.

**(c)** Give *two* disadvantages of using biofuels.

**13. (a)** Copy the table below. Sort the following resources and place one into each empty cell in the table.

**coal** **bamboo** **wind** **rock**

| | | Type of resource | |
|---|---|---|---|
| | | **renewable** | **non-renewable** |
| **How used** | **energy** | | |
| | **other** | | |

**(b)** Using examples from the table, describe *two* reasons why it is preferable to use a renewable resource rather than a non-renewable resource as a supply of energy.

**(c)** Using examples from the table, suggest what the 'other' uses for the renewable and non-renewable resources could be.

**14.** Trees are important natural resources because they use photosynthesis. The word equation for this reaction is:

$$\text{carbon dioxide + water} \xrightarrow{\text{sunlight}} \text{glucose + oxygen}$$

**(a)** Sunlight is needed to make this reaction happen. Is sunlight a renewable or non-renewable resource? Explain your answer.

**(b)** Photosynthesis uses up carbon dioxide and produces oxygen. Explain how photosynthesis can affect the mixture of gases in the atmosphere.

## Challenge questions

Scientists are concerned that human activities are speeding up the process of global warming.

**15.** Define the term 'global warming'.

**16.** Explain how the cutting down of large areas of forest can affect global warming.

12

# Chapter 12

## Earth in space

### What's it all about?

As well as the Sun and Earth and the other planets, our Solar System contains hundreds of thousands of rocky objects called asteroids. Investigating asteroids should give scientists evidence of how the Solar System formed. But our Sun is just one of about 250 000 000 000 (250 billion) stars in our galaxy, the Milky Way. If you look up at night in a place with few city lights, you can see the Milky Way as a ribbon stretching across the sky. Even if only a small percentage of those stars have their own 'solar systems' of planets, that would still mean many hundreds of millions of planets are out there!

You will learn about:

- Asteroids and how they were formed
- Our galaxy, the Milky Way, and what it contains

You will build your skills in:

- Describing what an analogy is and how it can be used
- Using an analogy to explain the problems involved with making an estimate of the number of stars in the Milky Way
- Describing how scientific hypotheses can be supported by evidence
- Presenting and interpreting observations and measurements

Chapter 12 . Topic 1

# Asteroids

You will learn:

- To describe what asteroids are
- To describe how asteroids form
- To describe how scientific hypotheses can be supported by evidence
- To present observations and measurements appropriately

## Starting point

| You should know that... | You should be able to... |
|---|---|
| The Sun and planets formed from gas and dust drawn together by the force of gravity | Identify patterns in data |
| The force of gravity holds planets in orbit around the Sun, and moons in orbit around planets | Evaluate a range of secondary information sources for their relevance |

## Rocks orbiting the Sun

When our Solar System formed billions of years ago, gas and dust clumped together due to the gravitational force to form the Sun and planets. During this process, many smaller clumps formed pieces of rock that continued to orbit the Sun without being pulled into the planets. They are not big enough to be planets and they don't have an atmosphere. These are **asteroids**. Most asteroids are made of rock but some contain large amounts of different metals.

As equipment to detect orbiting objects has improved, people have recorded more and more asteroids and their orbits. As of September 2019, the US space agency NASA has records of nearly 800 000 asteroids. By the time you read this, more will have been discovered!

There is a huge range of sizes of asteroids, from under 10 m across to over 530 km in diameter (figure 12.1). The smaller asteroids tend to have irregular (lumpy) shapes; the bigger asteroids have enough mass that, over time, they have become more rounded due to the effects of gravitational forces.

### Key term

**asteroid**: object composed of rock that is too small to be a planet and orbits the Sun.

a 

b 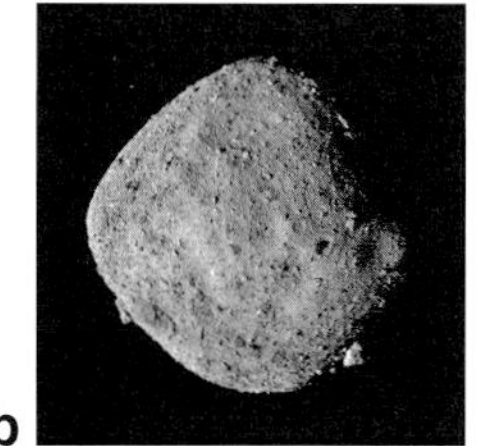

c 

**12.1** *Asteroids.* ***a*** *Ceres is the largest asteroid between the orbits of Mars and Jupiter. It is 945 km in diameter and very nearly spheroidal (meaning shaped like a ball).* ***b*** *Bennu is about 525 m in diameter and has an unusual shape.* ***c*** *Itokawa is only about 313 m in diameter but has an even more unusual shape.*

To give a sense of scale, if all the asteroids found so far were combined, the resulting object would still be smaller than Earth's Moon.

Ceres (figure 12.1a) is one of the largest asteroids at 945 km in diameter. It is classified as a dwarf planet, like Pluto, and orbits the Sun between the orbits of Mars and Jupiter. Bennu (figure 12.1b) is much smaller. It is thought to have formed between Mars and Jupiter but has drifted slowly nearer to Earth.

Bennu is currently being orbited by the space probe OSIRIS-REx, which will land on Bennu in late 2020 and collect a rock sample to bring back to Earth for scientists to analyse. Itokawa (figure 12.1c) is even smaller and was the first asteroid to be visited by a space probe. The Japanese probe Hayabusa returned a sample to Earth in 2010.

Most asteroids orbit the Sun in an area between the orbits of the planets Mars and Jupiter, forming an **asteroid belt**. Scientists believe a planet may have started to form in this area as the Solar System developed, but Jupiter became so large that its gravitational attraction pulled the forming planet apart, leaving the asteroids behind.

There are also two groups of asteroids called the **Trojans,** which are pulled around the Sun in the same orbit as Jupiter. Many other stars have been observed to have their own 'solar systems' containing planets. We call these **planetary systems.** Astronomers think that planetary systems form in the same way as our Solar System, meaning that we would expect there to be asteroids in other planetary systems.

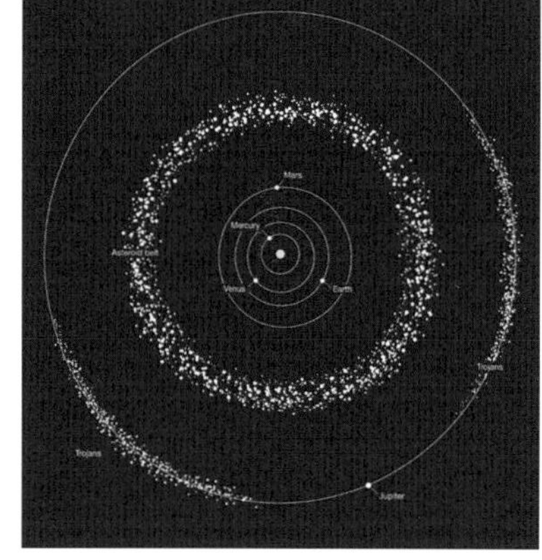

**12.2** *The asteroid belt and the Trojans.*

### Key terms

**asteroid belt**: area of space between the orbits of Mars and Jupiter that contains most of the asteroids in the Solar System.

**planetary system**: arrangement of planets and asteroids that forms around and orbits a star.

**Trojans**: asteroids that follow the same orbit as the planet Jupiter.

### Making links

How important are asteroids in our understanding of the Solar System? The ideas of how the Solar System formed are partly based on observations of asteroids. They are thought to be a stage in the formation of a planet, after gas and dust has clumped together, but before a full planet is formed (Stage 7 Chapter 11).

Suggest how long ago asteroids formed.

## Activity 12.1: Finding out about Psyche

Produce an information poster about the asteroid called Psyche. Include your answers to the following questions in your poster.

**A1** Where can Psyche be found in the Solar System?

**A2** What is the diameter of Psyche?

**A3** How long does it take for Psyche to orbit the Sun?

**A4** Psyche is different from many other asteroids because it appears to be made mostly from metal. Which are the two metals Psyche is made from?

**A5** There is a plan to launch a space probe to investigate Psyche. When will this probe be launched and how long will it take to reach the asteroid?

**A6** Challenge Scientists think that Psyche may have formed from the core of a planet that formed with the rest of the Solar System but was pulled apart. Suggest what is similar about Psyche and Earth. How might investigating Psyche help us learn more about the Earth?

1. Complete the sentences using the words from the box.

| rock | gas | Earth | Mars | Jupiter | Saturn |
|---|---|---|---|---|---|

Asteroids are mostly made from ___________ .

Most asteroids in the Solar System are found between the orbits of ___________ and ___________.

2. Describe how most asteroids formed.

3. Figure 12.3 shows part of our Solar System (**not** to scale).

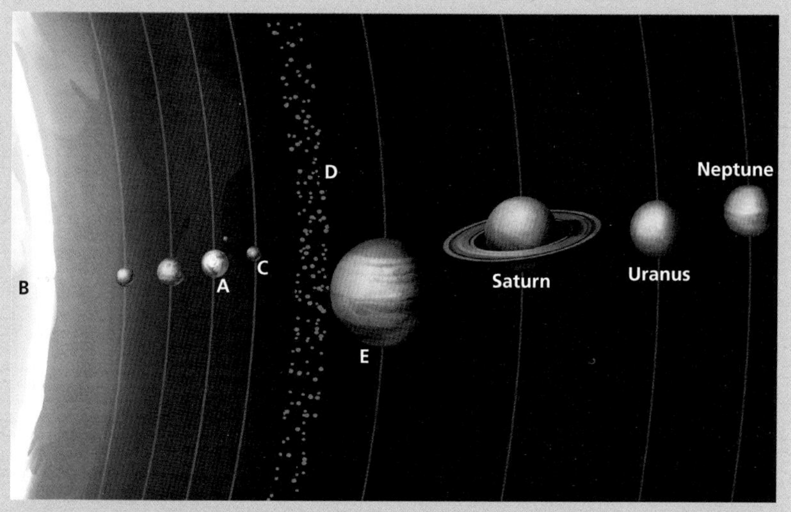


12.3

Copy and complete the labels by naming the objects found at points **A–E**.

**A** ________________

**B** ________________

**C** ________________

**D** ________________

**E** ________________

**4** Figure 12.4 shows two objects found in the Solar System. One is a planet (Mars) and one is a large asteroid (Ceres).

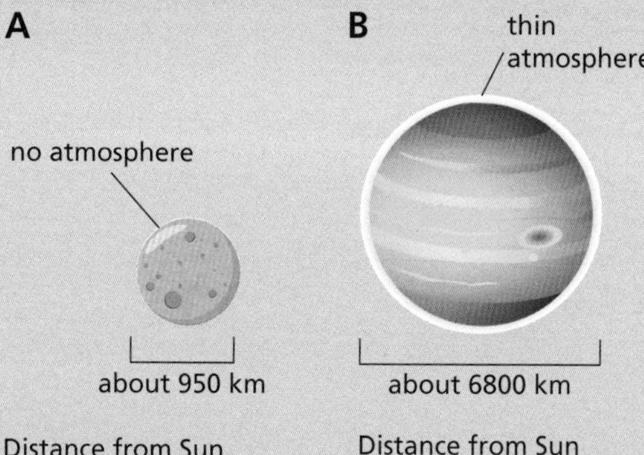


12.4

a) Use the evidence in the diagram to determine which object is Mars and which is Ceres.

**A** is ______________ .

**B** is ______________ .

b) Explain your answer to part **a** by describing *three* pieces of evidence you used.

**5** Evidence for asteroids can be found naturally on the surface of Earth.

a) Describe how this evidence can arrive on Earth.

b) Even though this evidence is available, scientists send space probes to collect samples of asteroids still in orbit. Suggest *three* reasons why these samples could be more useful than the evidence found naturally on Earth.

## Key facts:

- ✔ Asteroids in our Solar System are rocky objects that formed at the same time as the Solar System formed.
- ✔ Asteroids are also thought to form in other planetary systems around stars other than our Sun.
- ✔ Asteroids are smaller than planets and do not have an atmosphere.

## Check your skills progress:

- I can describe evidence for the formation of asteroids.
- I can research and present information about asteroids appropriately.

Chapter 12 . Topic 2

# Stars and galaxies

You will learn:

- To describe our galaxy, the Milky Way
- To understand how the stars, dust and gas in the Milky Way orbit the centre of the galaxy
- To explain an analogy and how to use it as a model
- To use analogies
- To discuss how scientific knowledge is developed
- To describe the application of science in society, industry and research
- To describe how scientific progress is made through individuals and collaboration

## Starting point

| You should know that... | You should be able to... |
|---|---|
| The force of gravity causes all objects to attract each other due to their mass | Describe the strengths and limitations of a model |
| The Sun and planets formed from gas and dust drawn together by the force of gravity | Record measurements in appropriate tables |

## The Milky Way

When you look up at the sky at night time, you can see many bright points of light. If you look using a telescope, some of those points of light look like discs – these are the other planets in our Solar System, reflecting the light of the Sun. Their positions in the sky change as they orbit the Sun. Some other points of light appear to move quite quickly – these are artificial satellites that humans have placed into orbit around the Earth.

All the other points of light are much further away, and every one is a star producing its own light. Human eyes are only sensitive enough to see about 5000 stars as separate points of light. People in ancient times thought the stars were fixed on a surface they called the 'firmament', which rotated around the Earth.

We know now that the stars are much further away, and that the Earth orbits the Sun. The Sun is one of many stars that form a disc-shaped 'cloud' called a **galaxy**. If we use modern telescopes, we see many more stars in our galaxy. Scientists estimate that there are about 250 billion (250 000 000 000) stars in our galaxy.

What would our galaxy, the **Milky Way**, look like if we could travel away from it and look back at it? Figure 12.5 shows what scientists think we would see.

### Key term

**galaxy:** collection of hundreds of millions of stars and their planetary systems, plus dust and gas, all orbiting a central point.

### Key term

**Milky Way:** the galaxy containing our Sun and Solar System.

**12.5** *Computer model of the Milky Way galaxy based on collected observations of its stars from Earth.*

All the stars in the Milky Way orbit a central point. At this point is an object that contains so much mass, the force due to its gravity pulls all the billions of stars into orbits. This object is called a **black hole**, because even light is pulled into it. Scientists think a black hole is what remains after a star becomes so large that it collapses in on itself due to the effect of gravity.

What this means is that, in the same way as the planets including Earth orbit our Sun, so our Sun orbits the centre of the Milky Way. The orbit is so large that scientists estimate our Sun takes 250 million years to complete one orbit.

### Key term

**black hole:** a collapsed star with a very high mass, found at the centre of most galaxies.

## Activity 12.2: Observing the Milky Way

Either as a class with your teacher, or with your family, read through the questions for this activity, then go outside on a clear night and observe the stars above you. You need to wait outside for at least ten minutes to allow your eyes to adjust to the dark.

Look upwards and observe. You need to remember the answers to these three questions.

**A1** Can you see lots of stars, or are there many lights of a town or city nearby creating 'light pollution' that make it hard to see many stars?

**A2** If it is dark enough, notice how the stars can appear to be arranged in patterns. In ancient civilisations, these apparent patterns were thought to represent animals (such as the Great Bear and Taurus, the bull) and characters from myths and legends (such as Orion the Hunter and Gemini, the twins). Different patterns are seen depending on the time of year and where you are located. Taurus, Orion and Gemini are all visible from both northern and southern hemispheres, depending on the time of year. Try to remember a pattern and draw it when you get inside.

**A3** If it is really dark, you may be able to see part of the structure of the Milky Way. It looks like a pale 'ribbon' running through the sky (see the photo at the start of this chapter). Can you see the Milky Way?

1. How many times does the Earth orbit the Sun in the time it takes for the Sun to orbit the centre of the Milky Way?

2. Explain how we can find evidence that our Sun is part of the Milky Way by observing the night sky.

3. Describe the object that can be found at the centre of the Milky Way. Include in your answer:
   - what it is called and why
   - the effect it has on all the stars in the Milky Way
   - how it was formed.

## Measuring distance

In our Solar System, we know that Earth is just under 150 000 000 km from the Sun. This is a large distance. To make it easier to talk about distances in the Solar System, we say that the Earth to Sun distance is 1 **astronomical unit** (abbreviated as AU). For example, on average Jupiter is 5.2 AU away from the Sun.

The scale of the Milky Way is so much larger than our Solar System that we need an even bigger measurement of distance. This is called the **light-year** and it is equal to the distance that light travels in space in one year.

The diameter of the Milky Way is about 105 000 light-years. Another way of describing this is to say it would take 105 000 years for a ray of light to cross from one side of the Milky Way to the other. The next nearest large galaxy, called Andromeda, is about 2.5 million light-years away.

### Key terms

**astronomical unit**: a unit of distance equivalent to the distance between Earth and the Sun, just under 150 million km.

**light-year**: a unit of distance equal to the distance that light travels in one year, about $9.5 \times 10^{12}$ km.

### Activity 12.3: Measuring very large distances

Table 12.1 shows a number of different, very large distances. They are shown in scientific notation, showing powers of 10. For example, $100 = 1.00 \times 10^2$ (because $100 = (10)^2$), and $0.001 = 1 \times 10^{-3}$.

Use a calculator to answer the questions, using scientific notation. Some numbers are very large, so do not worry if you run out of time; the important thing is to get practice in using your calculator with large values.

| Distance | Unit of distance | | |
| --- | --- | --- | --- |
| | kilometres (km) | astronomical units (AU) | light-year (ly) |
| travelled by light in one year | $9.5 \times 10^{12}$ | | |
| from Earth to Sun | $1.5 \times 10^{8}$ | 1.0 | $1.6 \times 10^{-5}$ |
| from Earth to next nearest star, Proxima Centauri | | | 4.2 |
| from Sun to centre of Milky Way | | | $2.6 \times 10^{4}$ |
| from Sun to next nearest galaxy, Andromeda | | | $2.5 \times 10^{6}$ |

***Table 12.1*** *Astronomical distances.*

**A1** Calculate the number of astronomical units in a light-year.

**A2** Complete the first three rows of the table.

**A3** Challenge Complete as much of the last two rows of the table as you can.

**4** Why do astronomers use astronomical units and light-years to describe distances?

**5** Calculate how many times larger the distance from Earth to Andromeda is compared to the diameter of the Milky Way.

## Other planetary systems, dust and gas

Astronomers have found evidence from many other stars in the Milky Way that they are orbited by planets. This suggests that a significant percentage of the stars in the Milky Way have their own planetary systems. It is reasonable to think that most other galaxies are similar to ours in terms of containing many planetary systems.

Even though there are very many stars and probably many planetary systems in galaxies, the main part of a galaxy's mass appears to be **interstellar gas** and **interstellar dust**. 'Interstellar' means 'between the stars'.

- Most of the gas is hydrogen, the simplest element. Hydrogen was the main part of the gas that formed our Solar System. It is used as a fuel for nuclear reactions in most stars.
- Most of the rest of the gas is helium, the next simplest element. Helium forms as part of the nuclear reactions in stars.

### Key terms

**interstellar dust**: tiny particles of materials outside stars and solar systems, containing mostly carbon, silicon and oxygen.

**interstellar gas**: gases found between stars and through space, containing mostly hydrogen and helium.

- The dust contains tiny particles of carbon, silicon and oxygen. Scientists believe these are produced and scattered through space when an ageing star explodes.

We can detect hydrogen, helium and dust in clouds that are closer to stars because they reflect energy that the stars produce. These clouds are important, because new stars form from these as the gases and dust pull together because of the force of gravity.

Hydrogen gas and other simple particles (such as electrons and protons) are thought by scientists to have formed very early on when the Universe began. As the forces due to gravity caused the gas to form the first stars, nuclear reactions in the stars produced elements with atoms larger than hydrogen, including helium.

When stars collapse or explode as substances to fuel their nuclear reactions run out, even larger atoms are formed such as those of metals like iron and nickel, and non-metals like carbon. These larger atoms form substances that are found in dust and rocks.

The stages in order can be summarised:

1. simple gas and dust gather together due to gravity, to form:
    - clumps of rock and some metals (producing asteroids)
    - larger clumps of rock with some gas (producing rocky planets)
    - larger clumps of gas (producing gas planets)
    - very large clumps of gas that start nuclear reactions (producing stars)
2. when the nuclear fuel runs out, stars collapse or explode…
3. …producing clouds of gas and dust.

**12.6** *One of many enormous clouds of gas and dust in the Milky Way galaxy. The hydrogen gas reflects light with a pink colour and the darker areas are formed from interstellar dust.*

**6** Describe the difference between interstellar dust and interstellar gas.

**7** Explain how astronomers are able to observe interstellar gas.

**8** The formation of interstellar dust and gas, stars and planetary systems (including asteroids and planets) is part of a cycle. Describe this cycle.

## Science in context: Finding exoplanets

An **exoplanet** is a planet that orbits a star other than our own Sun. ('Exo' means 'outside'.) Because they are so far away and do not produce their own light, it is not possible to see these exoplanets using telescopes that detect light. However, because of the way gravitational force works, an exoplanet also pulls on its 'parent' star. Large exoplanets cause the stars to 'wobble'. Astronomers can measure these wobbles using telescopes that detect other kinds of electromagnetic waves, such as radio waves.

**12.7** *Radio telescopes contain large dishes to collect radio waves.*

Often, many radio telescope dishes are arranged in patterns, and the signals they detect are added together. Having larger dishes and bigger numbers of dishes spread out over a larger area greatly improves the data collected in three ways:

- it increases the amount of detail that can be detected
- it means objects that are further away can be detected
- it means less bright objects can be detected.

Scientists from many countries around the world now link their telescopes together, so they can all observe the same object at the same time. The size of the telescope is effectively the size of Earth! This requires close collaboration on technology and sharing data.

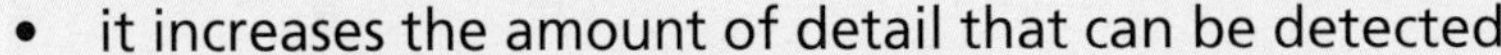

## Other galaxies

How can scientists know so much about the shape and size of the Milky Way, when Earth is inside the galaxy, so it is not actually possible to see the shape?

Astronomers estimate there are at least 100 billion galaxies in the Universe of many different shapes and sizes. Astronomers have observed many of the different shapes of galaxies we can see and compared them with measurements of the stars in our galaxy. These comparisons tell them the Milky Way's shape and size.

### Key term

**exoplanet:** planet orbiting a star other than our Sun.

**12.8** *Different shapes of galaxies.*

Astronomers think that there are 250 billion stars in the Milky Way, but this is only an **estimate**. In fact, it is very difficult to make an estimate of this number, because a large part of the Milky Way is hidden from view. When we look towards the centre of the galaxy, the stars, dust and gas between us and the centre block our view of the other half of the galaxy. A more complete description of the estimate is to say:

250 billion stars ± 150 billion stars

### Key term

**estimate:** guess at a value, based on evidence that is explained.

## Activity 12.4: Making estimates

You will now make an estimate and describe how good your estimate is. At the end of this activity, you will look at the estimate of the number of stars in the Milky Way and decide how good an estimate that is.

12.9

1 Figure 12.9 shows a box full of coloured balls. How many balls can you count in 30 seconds?

2 Note that the box is divided into nine equally sized parts. Count the number in one of these parts and multiply this number by 9.

3 Steps 1 and 2 are two different methods of estimating the number of a large number of similar objects. Compare the two estimates. Which method seems easier?

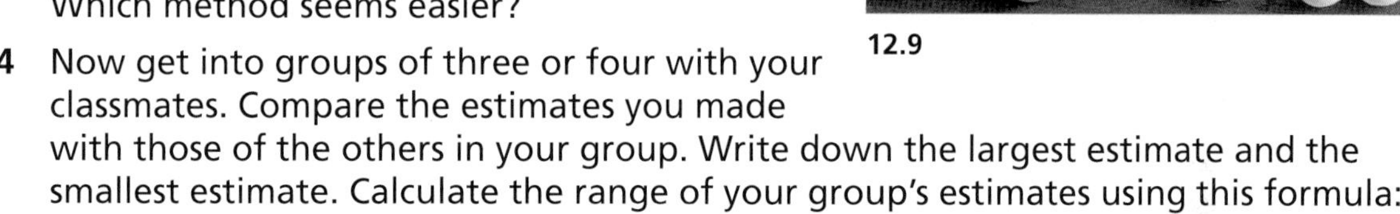

4 Now get into groups of three or four with your classmates. Compare the estimates you made with those of the others in your group. Write down the largest estimate and the smallest estimate. Calculate the range of your group's estimates using this formula:

**range** = largest estimate – smallest estimate

5 Add up all the estimates and divide by the number of estimates. This will give you the mean estimate:

$$\text{mean estimate} = \frac{\text{total of all estimates}}{\text{number of estimates}}$$

6 You can now state your estimate in the following way:

$$\text{estimate} = \text{mean estimate} \pm \frac{\text{range}}{2}$$

7 In Figure 12.9, you can only see the balls on top of the box. There is no way to know how deep the box is or how many balls are hidden from view. How accurate do you think your estimate is?

This is an **analogy** for the problem astronomers face in trying to estimate the number of stars in the galaxy. It helps explain why an estimate is difficult to make and why an estimate might not be very accurate.

**A1** Based on steps 4–6, explain the different parts of the astronomers' estimate of the number of stars in the Milky Way:

250 billion stars ± 150 billion stars

**A2** Describe how astronomers could use the method described in step 2 to make it easier to estimate the number of stars.

**A3** Determine the range of the astronomers' estimate.

**A4** Comment on what the range tells you about the astronomers' estimate. Is it very accurate?

**A5** Use step 7 to describe a possible reason for the accuracy of the astronomers' estimate.

**9** Describe how observing other galaxies can help astronomers understand more about our own galaxy.

**10** Describe an analogy for estimating the number of stars in the Milky Way.

**11** Astronomers have developed computer models of how the stars in a galaxy orbit the centre. These models can show in hours an idea of how a galaxy changes over hundreds of millions of years. Suggest how astronomers can test their models against evidence.

### Key terms

**analogy**: a type of model that compares something unfamiliar to something more familiar.

**range**: the difference between the smallest measurement and the largest measurement of the same quantity.

### Key facts:

- ✔ Our galaxy, the Milky Way, contains stars, planetary systems, interstellar gas and interstellar dust.
- ✔ Our Sun is one of hundreds of billions of stars that orbit the centre of the Milky Way.
- ✔ The stars in a galaxy orbit a central black hole, because of the effect of the force of gravity.

### Check your skills progress:

- I can describe how the Milky Way can be modelled using observations of other galaxies.
- I can explain what an analogy is.
- I can use an analogy to explain how astronomers can estimate the number of stars in the Milky Way.

# End of chapter review

## Quick questions

1. What is an asteroid made from?

   **(a)** mostly gas

   **(b)** mostly rock or metal

   **(c)** mostly dust

   **(d)** rock surrounded by an atmosphere

2. What object do asteroids in our Solar System orbit?

   **(a)** Jupiter

   **(b)** Earth

   **(c)** Mars

   **(d)** the Sun

3. How many stars are there in the Milky Way?

   **(a)** about 100 000 000

   **(b)** about 250 000 000

   **(c)** about 100 000 000 000

   **(d)** about 250 000 000 000

4. What is a light-year?

   **(a)** The time it takes Earth to orbit the Sun

   **(b)** The distance light travels in one year

   **(c)** The time it takes for light to cross the Milky Way

   **(d)** The distance to the nearest star from Earth (apart from the Sun)

5. Tick the statements that are true about asteroids.

   ☐ **a** Some asteroids have an atmosphere.

   ☐ **b** Fewer than 500 000 asteroids have been discovered.

   ☐ **c** Asteroids orbit the Sun.

   ☐ **d** All asteroids are too dark to be seen from Earth.

   ☐ **e** Pieces of some asteroids have been found on the surface of Earth.

6. (a) Describe *two* differences between planetary systems and galaxies.

   (b) Describe *one* similarity between planetary systems and galaxies.

## Connect your understanding

7. (a) Explain why astronomers use the unit of the light-year to measure distances within and between galaxies.

   (b) Suggest why it would be very difficult for humans to travel to the nearest large galaxy, Andromeda.

8. (a) Name the force that is the reason why stars orbit the centre of a galaxy.

   (b) Explain why it is difficult for astronomers to observe all of our own galaxy, the Milky Way.

   (c) Describe how astronomers can still find evidence to support their ideas about the size, shape and behaviour of the Milky Way.

9. This short article describes the asteroid Bennu. Read the article, then answer the questions that follow.

   Bennu is a small asteroid with an orbit that brings it much closer to Earth than the main asteroid belt. It is very old, possibly older than the planets in the Solar System. Scientists think it may contain substances that were needed on Earth for life to begin. Bennu also contains many valuable materials, such as rocks rich in iron and aluminium, and precious metals such as platinum. A space probe called OSIRIS-Rex has been investigating Bennu and will collect a sample to bring back to Earth.

   (a) Suggest why it is easier to send a space probe to Bennu than most other asteroids.

   (b) Describe how Bennu could help us to understand how the Solar System formed.

   (c) Suggest why biologists are interested in examining the sample that will be brought back from Bennu.

   (d) Suggest how Bennu could help address problems with the supply of non-renewable resources on Earth.

10. (a) Name the type of object found at the centre of a galaxy.

    (b) List *five* types of objects that can be found in galaxies.

    (c) Explain the importance of the force due to gravity in:

      (i) the formation and behaviour of asteroids

      (ii) the formation and behaviour of galaxies.

**11.** Sometimes small pieces of rock from space are pulled towards the Earth. These rock pieces travel at high speed and mostly burn up in the atmosphere. Occasionally a piece will reach the surface. We call these surviving rock pieces meteorites.

**(a)** These small pieces of rock break off from larger objects made from rock, which orbit the Sun. Name these larger objects.

**(b)** Name the force that pulls these rock pieces towards Earth.

**(c)** **Challenge** Suggest an explanation for the rock pieces burning up in the atmosphere.

**(d)** Explain how scientists can use these meteorites as evidence for the formation of the Solar System.

## End of stage review

1. Select the *two* correct statements.

   a Climate describes atmospheric conditions over short times and local areas.

   b Climate describes atmospheric conditions over long times and large areas.

   c Weather describes atmospheric conditions over short times and local areas.

   d Weather describes atmospheric conditions over long times and large areas.

2. The graph shows the change in sea level on Earth since 1880.

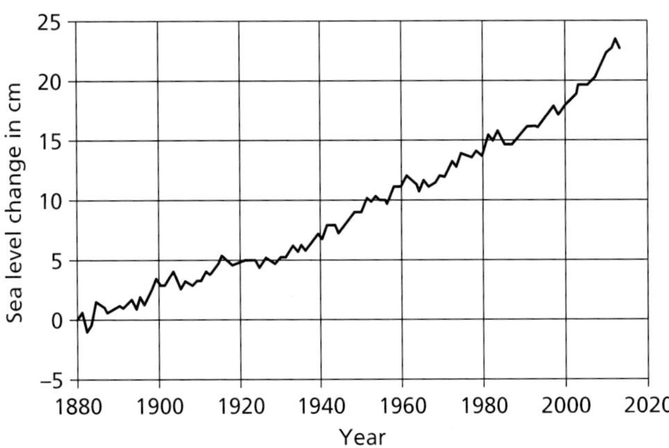


   **(a)** Describe the trend in sea level shown by the graph.

   **(b)** Describe how changes in sea level are linked to changes in the amount of ice at Earth's poles and in glaciers.

   **(c)** Explain how this graph provides evidence of climate change.

3. **(a)** State what is meant by the term 'renewable resource'.

   **(b)** Give *two* examples of renewable resources that are used to generate electricity.

   **(c)** **(i)** Predict whether the percentage of electricity generated globally from renewable resources will increase or decrease over the next 30 years.

   **(ii)** Give a reason for your answer.

4. **(a)** Describe *two* characteristics of asteroids.

   **(b)** Describe *two* ways in which humans have been able to obtain pieces of asteroids to examine.

**5.** The diagram shows what astronomers think our galaxy, the Milky Way, looks like.

**(a)** Name *four* types of objects or materials that make up most of the mass in the Milky Way.

**(b)** Describe what causes these objects and materials to form the spiral shape shown in the diagram.

# Periodic Table

**Key**

relative atomic mass
**atomic symbol**
name
atomic (proton) number

| 1 | 2 | | | | | | | | | | | 3 | 4 | 5 | 6 | 7 | 0 |
|---|---|---|---|---|---|---|---|---|---|---|---|---|---|---|---|---|---|
| | | | | | | | 1<br>**H**<br>hydrogen<br>1 | | | | | | | | | | 4<br>**He**<br>helium<br>2 |
| 7<br>**Li**<br>lithium<br>3 | 9<br>**Be**<br>beryllium<br>4 | | | | | | | | | | | 11<br>**B**<br>boron<br>5 | 12<br>**C**<br>carbon<br>6 | 14<br>**N**<br>nitrogen<br>7 | 16<br>**O**<br>oxygen<br>8 | 19<br>**F**<br>fluorine<br>9 | 20<br>**Ne**<br>neon<br>10 |
| 23<br>**Na**<br>sodium<br>11 | 24<br>**Mg**<br>magnesium<br>12 | | | | | | | | | | | 27<br>***Al***<br>aluminium<br>13 | 28<br>**Si**<br>silicon<br>14 | 31<br>**P**<br>phosphorus<br>15 | 32<br>**S**<br>sulfur<br>16 | 35.5<br>***Cl***<br>chlorine<br>17 | 40<br>**Ar**<br>argon<br>18 |
| 39<br>**K**<br>potassium<br>19 | 40<br>**Ca**<br>calcium<br>20 | 45<br>**Sc**<br>scandium<br>21 | 48<br>**Ti**<br>titanium<br>22 | 51<br>**V**<br>vanadium<br>23 | 52<br>**Cr**<br>chromium<br>24 | 55<br>**Mn**<br>manganese<br>25 | 56<br>**Fe**<br>iron<br>26 | 59<br>**Co**<br>cobalt<br>27 | 59<br>**Ni**<br>nickel<br>28 | 63.5<br>**Cu**<br>copper<br>29 | 65<br>**Zn**<br>zinc<br>30 | 70<br>**Ga**<br>gallium<br>31 | 73<br>**Ge**<br>germanium<br>32 | 75<br>**As**<br>arsenic<br>33 | 79<br>**Se**<br>selenium<br>34 | 80<br>**Br**<br>bromine<br>35 | 84<br>**Kr**<br>krypton<br>36 |
| 85<br>**Rb**<br>rubidium<br>37 | 88<br>**Sr**<br>strontium<br>38 | 89<br>**Y**<br>yttrium<br>39 | 91<br>**Zr**<br>zirconium<br>40 | 93<br>**Nb**<br>niobium<br>41 | 96<br>**Mo**<br>molybdenum<br>42 | [98]<br>**Tc**<br>technetium<br>43 | 101<br>**Ru**<br>ruthenium<br>44 | 103<br>**Rh**<br>rhodium<br>45 | 106<br>**Pd**<br>palladium<br>46 | 108<br>**Ag**<br>silver<br>47 | 112<br>**Cd**<br>cadmium<br>48 | 115<br>**In**<br>indium<br>49 | 119<br>**Sn**<br>tin<br>50 | 122<br>**Sb**<br>antimony<br>51 | 128<br>**Te**<br>tellurium<br>52 | 127<br>**I**<br>iodine<br>53 | 131<br>**Xe**<br>xenon<br>54 |
| 133<br>**Cs**<br>caesium<br>55 | 137<br>**Ba**<br>barium<br>56 | 139<br>**La***<br>lanthanum<br>57 | 178<br>**Hf**<br>hafnium<br>72 | 181<br>**Ta**<br>tantalum<br>73 | 184<br>**W**<br>tungsten<br>74 | 186<br>**Re**<br>rhenium<br>75 | 190<br>**Os**<br>osmium<br>76 | 192<br>**Ir**<br>iridium<br>77 | 195<br>**Pt**<br>platinum<br>78 | 197<br>**Au**<br>gold<br>79 | 201<br>**Hg**<br>mercury<br>80 | 204<br>***Tl***<br>thallium<br>81 | 207<br>**Pb**<br>lead<br>82 | 209<br>**Bi**<br>bismuth<br>83 | **Po**<br>polonium<br>84 | **At**<br>astatine<br>85 | **Rn**<br>radon<br>86 |
| **Fr**<br>francium<br>87 | **Ra**<br>radium<br>88 | **Ac****<br>actinium<br>89 | **Rf**<br>rutherfordium<br>104 | **Db**<br>dubnium<br>105 | **Sg**<br>seaborgium<br>106 | **Bh**<br>bohrium<br>107 | **Hs**<br>hassium<br>108 | **Mt**<br>meitnerium<br>109 | **Ds**<br>darmstadtium<br>110 | **Rg**<br>roentgenium<br>111 | | | | | | | |

| |
|---|
| **La**<br>lathanoids |
| **Ac**<br>actinoids |

Elements 1 to 92 are naturally occurring elements on Earth. Elements 93 and above are man-made.

# Glossary

**absorption**: the way in which an object takes in the energy reaching its surface.

**accurate**: an accurate result is one that is close to the real answer.

**addictive**: substance that makes people feel that they must have it.

**aerobic respiration**: respiration that requires oxygen to release energy from glucose.

**alkaline**: having the properties of an alkali.

**alloy**: a mixture of metals with other elements.

**alpha particle**: particle with a positive charge given out by some radioactive elements – it is smaller than an atom.

**alveolus** (plural alveoli): tiny, pocket-shaped structure in lungs where gas exchange happens.

**analogy**: a type of model that compares something unfamiliar to something more familiar.

**angle of incidence**: this is the angle between the incident ray and the normal.

**angle of reflection**: this is the angle between the reflected ray and the normal.

**angle of refraction**: this is the angle between the refracted ray and the normal.

**anomalous results**: results which don't fit the pattern of the other results obtained.

**antagonistic pair**: two muscles that pull a bone in opposite directions.

**apparent depth**: how deep something appears to be.

**asteroid**: object composed of rock that is too small to be a planet and orbits the Sun.

**asteroid belt**: area of space between the orbits of Mars and Jupiter that contains most of the asteroids in the Solar System.

**astronomical unit**: a unit of distance equivalent to the distance between Earth and the Sun, just under 150 million km.

**attract**: pull closer together.

**average**: the mean average of a set of numbers is found by: total of all the numbers added together/how many numbers there are

**balanced diet**: eating many different foods to get the correct amounts of nutrients.

**balanced forces**: when the resultant force is zero.

**ball and socket joint**: joint where a ball-shaped piece of bone fits into a socket made by other bones.

**bioaccumulation**: build-up of a substance in an organism because the substance cannot be broken down and is not excreted.

**biofuels**: fuels made from renewable sources such as crops.

**bioplastic**: a material that can be used in the same way as plastics produced from oil, but which is produced using renewable sources.

**black hole**: a collapsed star with a very high mass, found at the centre of most galaxies.

**blood**: liquid tissue that carries substances around the body.

**blood vessels**: tube-shaped organs that carry blood around the body.

**breathing**: movements of muscles in your respiratory system that cause air to move in and out of your lungs.

**breathing rate**: the number of times you inhale and exhale in one minute.

**bronchioles**: small tubes leading from the bronchus in a lung.

**bronchus** (plural bronchi): large tube leading from the trachea into a lung.

**cancer**: when cells in a tissue start to make many copies of themselves very quickly.

**capillary**: tiny blood vessel found in all the tissues of your body.

# Glossary

**carbohydrate**: nutrient needed for energy. Examples include starch and sugars (such as glucose).

**carbonate**: compound that contains carbon, oxygen and another element, for example, calcium carbonate ($CaCO_3$).

**carnivore**: animal that eats other animals.

**chemical reaction**: a change in which new substances are produced.

**chest**: area inside the body between the ribcage, neck, backbone and diaphragm.

**chloride**: compound that is formed when chlorine reacts with another element, for example, sodium chloride (NaCl).

**chromatogram**: the pattern of spots produced during chromatography.

**chromatography**: a technique used to separate soluble substances (usually coloured dyes or inks).

**cilia**: waving strands that stick out of some cells.

**ciliated epithelial cell**: specialised cell with waving cilia to sweep mucus along.

**circulation**: movement of blood around the body.

**circulatory system**: group of organs that move blood around the body.

**climate**: what happens in the air over very large areas for long periods of time (many years).

**climate change**: changes to the average long-term temperature and weather patterns of the Earth.

**combustion**: reaction of a substance with oxygen.

**competition**: a struggle between some organisms for the same resources.

**concentrated solution**: a solution that has a large number of solute particles dissolved in a small volume of solvent.

**concentration**: a measurement of how many particles of a certain type there are in a volume of liquid or gas.

**constipation**: when your intestines become blocked.

**consumer**: animal that eats other living things.

**contract** (muscle): when muscle tissue gets shorter and fatter it contracts.

**control variable**: variable that you keep the same during an investigation.

**core**: piece of metal (usually iron) that a coil of wire is wound around to increase the strength of the magnetic field.

**decomposer**: microorganism that causes decay.

**dependent variable**: variable you decide to measure in an experiment.

**diaphragm**: organ that helps with breathing

**diet**: what you normally eat and drink.

**diffraction grating**: transparent piece of glass or plastic which has many lines drawn onto it. Light can pass through the spaces between the lines.

**diffusion**: the spreading out of particles from where there are many (high concentration) to where there are fewer (lower concentration).

**dilute solution**: a solution that has a small number of solute particles dissolved in a large volume of solvent.

**dispersion**: the splitting of white light into a spectrum of colours.

**dissolve**: when a soluble substance becomes a solution.

**ecosystem**: all the organisms and the physical factors in an area.

**effort**: the force that you put in to move the load.

**electromagnet**: magnet that can be switched on or off using an electric current.

**electrons**: very small negatively charged particles in an atom.

**endothermic**: reaction or process in which energy is transferred from the surroundings, usually by heating, causing the temperature of the surroundings to decrease.

**environment**: the other organisms and physical factors around a certain organism.

**estimate**: guess at a value, based on evidence that is explained.

**evaluating an investigation**: describing what you would do differently in an experiment if you repeated it and explaining why you would do those things differently.

**exhale**: breathing out.

**exoplanet**: planet orbiting a star other than our Sun.

**exothermic**: reaction or process in which energy is transferred to the surroundings, usually by heating, causing the temperature of the surroundings to increase.

**extinction**: when a species dies out completely.

**faeces**: solid waste material produced by humans and other animals.

**fats**: nutrients needed by your body to store energy.

**fibre**: food substance that cannot be digested but which helps to keep your intestines healthy.

**filter**: a colour filter will only allow light of its own colour to pass through it.

**food chain**: diagram showing feeding relationships in a habitat – each species is a food source for the species at the next level up.

**food web**: diagram to show how food chains interconnect in a habitat.

**fossil fuels**: non-renewable resources that are formed from the remains of dead animals and plants.

**frequency**: the number of times something occurs.

**fuel**: substance that releases energy.

**galaxy**: collection of hundreds of millions of stars and their planetary systems, plus dust and gas, all orbiting a central point.

**gas exchange**: when two or more gases move from place to place in opposite directions.

**gas pressure**: the effect of the forces caused by collisions from gas particles on the walls of a container.

**glacial period**: time during an ice age when ice covers more of the Earth and the climate is cooler; it can last between 20 000 and 80 000 years.

**glacier**: slow-moving ice 'river' formed from snow squeezed together over more than one hundred years.

**global warming**: increasing temperatures on the Earth and in its atmosphere.

**gradient**: the gradient of a graph tells you how steep the line is.

**greenhouse gas**: gas in the atmosphere that reflects energy back to the surface, causing the temperature to rise.

**guard cell**: cell that helps form a stoma in a leaf, to allow gases in and out.

**habitat**: the place where an organism lives.

**haemoglobin**: substance that traps oxygen.

**hand lens**: another term for magnifying glass.

**hazard**: harm that something may cause.

# Glossary

**heart disease**: when the heart does not work well.

**herbivore**: animal that eats plants.

**high blood pressure**: when the pressure of blood inside your blood vessels is at risk of bursting them.

**hinge joint**: joint where two bones form a hinge.

**hydrocarbon**: substance made only of carbon and hydrogen.

**hydroelectric power**: electricity produced using energy from moving water as it is released from behind a dam.

**hydroxide**: compound that contains one atom each of oxygen and hydrogen bonded together; for example, potassium hydroxide (KOH).

**hypothesis** (plural hypotheses): a statement or claim that can be tested using experiments. It is proved or disproved by scientific enquiry.

**ice age**: period of time when Earth is several degrees Celsius colder than usual, due to changes in the Earth's orbit.

**ice age cycle**: cycle of Earth's climate changing from warm period to ice age and back again.

**ice core**: sample taken from deep in the ice and analysed scientifically.

**ice sheet**: large covering of ice at the North and South Poles of Earth.

**illegal drug**: drug that individual people are not allowed to buy or use. Different countries have different laws about drugs.

**incident ray**: this ray shows the light travelling towards the mirror.

**independent variable**: variable you decide to change in an experiment.

**inert**: a chemical that is unreactive.

**inhale**: breathing in.

**insoluble**: a substance that does not dissolve.

**intercostal muscles**: muscles that join the ribs together, and move the ribcage to change the volume of your chest during breathing.

**interglacial period**: time during an ice age when ice covers less of the Earth and the climate is warmer.

**interstellar dust**: tiny particles of materials outside stars and solar systems, containing mostly carbon, silicon and oxygen.

**interstellar gas**: gases found between stars and through space, containing mostly hydrogen and helium.

**invasive species**: a non-native species that damages an ecosystem.

**joint**: place in your skeleton where bones meet.

**joule**: unit used to measure energy.

**light-year**: a unit of distance equal to the distance that light travels in one year, about $9.5 \times 10^{12}$ km.

**line graph:** graph of two variables, both measured in numbers.

**line of best fit**: straight or curved line drawn through the middle of a set of points to show the pattern of data points.

**lipids**: another word for fats.

**load**: the force that you need to move.

**lungs**: organs that get oxygen into the blood and remove carbon dioxide.

**magnet**: an object, usually made from iron, nickel or cobalt, that has its atoms aligned so that the object has a magnetic field.

**magnetic field**: the region around a magnetic material in which a magnetic force acts.

**magnetic force**: force that occurs when a magnet attracts another object or repels another magnet.

**magnetic pole**: point on a magnet where the force is strongest.

**magnetise**: to make magnetic.

**magnetism**: property of some materials that gives rise to forces between these materials and magnets.

**magnifying glass**: used to make things appear bigger (magnify them).

**Milky Way**: the galaxy containing our Sun and Solar System.

**minerals**: nutrients that living organisms need in small amounts for health, growth and repair. Also called mineral salts.

**mixture**: two or more elements or compounds mixed together. They can easily be separated.

**model**: simple way of showing or explaining a complicated object or idea based on evidence.

**moment**: turning effect of a force – measured in newton-metres.

**mucus**: sticky liquid that traps particles.

**native species**: a species that naturally lives in an area.

**neutralisation**: chemical reaction between an acid and an alkali which produces a neutral solution.

**neutrons**: particles with no charge in the nucleus of an atom.

**nicotine**: addictive drug in tobacco smoke.

**non-native species**: a species that does not naturally live in a certain area.

**non-renewable resource**: resource that cannot be replaced easily and will run out one day.

**normal**: this is a line drawn at 90° to the mirror at the point where rays hit the mirror.

**normal force**: When an object touches a surface there is a force on the object at 90° to the surface (normal to the surface).

**nucleus**: the central part of an atom – contains protons and neutrons.

**nutrient**: a substance that an organism needs to stay healthy and survive.

**obesity**: being so overweight that your health is in danger.

**omnivore**: animal that eats both plants and animals.

**organ system**: group of organs working together.

**oxidation**: chemical reaction with oxygen to form a compound that contains oxygen.

**oxide**: compound that is formed when oxygen reacts with another element; for example magnesium oxide (MgO).

**pathogen**: a microorganism that causes a disease.

**permanent magnet**: object made from a magnetic material that retains its magnetism for a very long time.

**pharmaceutical drug**: drug used in healthcare to help the body fight a disease, or to relieve pain.

**physical factor**: non-living part of an environment (e.g. wind).

**pitfall trap**: jar buried in the ground to collect small animals that walk on the ground.

**pivot**: the point about which the force turns.

**plane mirror**: plane means flat, so a plane mirror is a flat mirror.

**planetary system**: arrangement of planets and asteroids that forms around and orbits a star.

**plasma**: liquid part of the blood.

# Glossary

**plastic materials**: materials derived (made) from oil and processed so they can be moulded.

**plum pudding model**: early model of the atom – a cloud of positive charge with electrons embedded in it.

**pooter**: device to suck small animals into a collecting jar without harming them.

**population**: the total number of individual organisms of one species living in a certain area.

**precipitate**: insoluble solid formed when soluble substances react together.

**precise**: how precise your measurement is depends on the measuring equipment and the smallest difference it can measure. The smaller the difference it can measure, the more precise the measuring equipment is.

**pressure**: the amount of force per unit of area – usually measured in newtons per square metre.

**primary colours**: red, blue and green. Mixing these colours of light together will make all other colours of light.

**primary consumer**: animal that eats plants (producers). These are the second trophic level in an ecosystem. For example, an antelope eats grass.

**prism**: transparent object that refracts light.

**producer**: organism that makes its own food, such as a plant.

**product**: substance made during a chemical reaction.

**proteins**: nutrients you need for growth and repair.

**protons**: positively charged particles in the nucleus of an atom.

**pure**: substance that contains only one element or compound.

**purity**: how much of a chemical is in a mixture.

**quadrat**: square frame used to take samples in a habitat.

**range**: the difference between the smallest measurement and the largest measurement of the same quantity.

**rate**: measurement of how quickly something happens.

**reactant**: substance that changes in a chemical reaction to form products.

**reactivity**: how likely it is that a substance will undergo a chemical reaction.

**reactivity series**: series of metals written in order from the most reactive to the least reactive.

**real depth**: how deep something really is.

**red blood cell**: cell that contains haemoglobin so it can carry oxygen.

**reflected ray**: this shows the light travelling away from the mirror after it has been reflected.

**refraction**: the bending of light when it enters a different medium.

**renewable resource**: resource that can be replaced easily in a short time.

**relax** (muscle): when muscle tissue stops contracting it relaxes.

**reliable**: measurements are reliable when repeated measurements give results that are very similar.

**repel**: push further apart.

**resource**: anything that is needed or used by an organism.

**respiration**: chemical process that happens inside cells to release energy.

**respiratory system**: group of organs that get oxygen into the blood and remove carbon dioxide. Also called the breathing system.

**resultant force**: shows the single total force acting on an object when all the forces acting on it are added up.

**rib**: bone that helps to protect your heart and lungs.

**ribcage**: all your ribs.

**risk**: chance of a hazard causing harm.

**risk assessment**: the process of identifying the hazards involved in a practical investigation and deciding how to control them.

**salt**: a type of compound that consists of metal atoms joined to non-metal atoms, e.g. sodium chloride.

**sample**: small portion of something, used to discover what the whole of the thing is like.

**scattering**: scattering happens when light is reflected from particles and uneven surfaces.

**secondary colours**: yellow, magenta and cyan.

**secondary consumer**: animal that eats a primary consumer. These are the third trophic level in an ecosystem. For example, a hyena eats an antelope.

**skeletal system**: group of organs that support the body and allow movement.

**skeleton**: another term for your skeletal system.

**skull**: a collection of bones that protect your brain.

**solar power**: electricity produced using energy from sunlight.

**solubility**: the mass of solute that will dissolve in a volume of solvent at a certain temperature.

**soluble**: substance that dissolves to form a solution.

**solute**: a substance that dissolves in the solvent, e.g. salt.

**solution**: a mixture of a soluble substance and a liquid.

**solvent**: a liquid that dissolves a soluble substance.

**speed**: how far something moves in a given time.

**stoma** (plural stomata): hole in a leaf formed between two guard cells.

**sugar**: soluble carbohydrate, which exists as small particles. Glucose is an example.

**sulfate**: compound that contains sulfur, oxygen and another element, for example copper sulfate ($CuSO_4$).

**surface area**: the area of a surface, measured in squared units such as square centimetres ($cm^2$).

**tar**: sticky black liquid found in cigarette smoke.

**thermal energy**: energy stored in an object due to its temperature.

**trachea**: tube-shaped organ that allows air to flow in and out of your lungs.

**trend**: pattern seen in data in which there is a change in a certain direction.

**Trojans**: asteroids that follow the same orbit as the planet Jupiter.

**trophic level**: level in an ecosystem. All producers are in the first trophic level. Energy passes from the lower to the higher trophic levels.

**tumour**: a lump of cancer cells.

**type 2 diabetes**: disease in which there is too much glucose in your blood, which can damage organs.

**unbalanced forces**: when there is a resultant force.

**Universal Indicator**: type of indicator which can change into a range of colours depending on whether the solution is acidic or alkaline and how strong it is.

**upthrust**: the upwards force from a liquid on an object in a liquid. Also applies to the upwards force on an object in a gas.

## Glossary

**variable**: something you can measure or observe.

**vertebra** (plural vertebrae): the bones in your back.

**vitamins**: nutrients you need in small amounts for health, growth and repair.

**volume**: how much space a substance takes up. Measured in $cm^3$.

**weather**: what happens in the air around a local area for a short time (hours or days).

**white blood cell**: cell that helps destroy microorganisms.

**word equation**: model showing what happens in a chemical reaction, with reactants on the left of an arrow and products on the right.

# Index


A

absorption 191, 267
accuracy 111, 132, 133, 267
addictive substances 41, 267
aerobic respiration 8–10, 267
air pollution 16, 172
air pressure 169
alcohol fuels 118
alkaline 120, 267
alloys 106, 267
alpha particles 80–81, 267
alveoli 10, 40, 267
analogy 79, 80, 167, 259, 260, 267
angle of incidence 180–181, 186, 267
angle of reflection 180–181, 267
angle of refraction 185–186, 267
anomalous results 138, 180, 267
antagonistic pair 19, 267
apparent depth 186–187, 267
arm muscles 19–20, 156
asteroid belt 249, 267
asteroids 247, 248–252, 267
astronomical units 255–256, 267
atmosphere 230–234
atoms 78
 rearranged in chemical reactions 103–104
 structure 79–82
attraction 82, 201–202, 267
average speed 137–140

B

balanced diet 28–36, 267
balanced forces 149–153, 267
balancing moments 157–160
ball and socket joints 19, 267
bar magnets 201, 202, 204–205, 206
batteries 123
bioaccumulation 60–63, 267
biofuels 240, 267
bioplastics 240, 243, 267
black holes 254, 267
blood 3–7, 89, 267
blood vessels 2, 3–4, 267
blood transfusions 5
bones 17–22
breathing 12–13, 267
breathing rate 12, 13, 267
bronchi 11, 267
bronchioles 11, 267

C

cancer 41–42, 267
capillaries 4, 10, 267
carbohydrates 8, 28, 29, 31–32, 35, 268
carbon dioxide 4, 9, 10, 89, 95, 230–233
carnivores 58, 268
chemical reactions 100–130, 268
 energy transfers in 115–118
 measuring temperature changes 109–112
 pure substances and mixtures 105–108
 reactivity series 119–125
 using word equations 101–104
choice chamber 51–52
chromatograms 83, 84–85, 268
chromatography 83–86, 268
cilia 11–12, 40, 268
ciliated epithelial cells 11–12, 268
circulation 3, 268
circulatory system 3, 268
climate 223–229, 268
 natural changes in 226–229
climate change 230–234, 268
cold packs 110, 111, 115
colour blindness 195
colour filters 192–194, 268
coloured light 189–195
combustion 232–233, 268
compasses 200, 202–207
competition 64–68, 268
concentrated solutions 92, 268
concentration 90–92, 170, 268
consumers 58–60, 268, 272, 273
contraction of muscles 12, 19–20, 268
control variable 6, 33, 34, 212, 268
core 209, 210–212, 268
cranes 160
current 209, 210–211

D

dependent variable 6, 33, 212, 268
depth 168
 real and apparent 186–187
diaphragm 11, 13, 14–15, 268
diet 28–36, 268
diffraction grating 189–190, 268
diffusion 9–10, 170–173, 268
 rate of 172–173
dilute acids 110, 121–122
dilute solutions 91, 92, 268
dispersion 189–190, 268
dissolving 83, 84, 90, 91, 105, 114, 268
 temperature changes during 111–112
distance
 astronomical distances 255–256
 measuring 133, 134–135
distance/time graphs 141–144
drugs 40–41, 95

# Index


E

Earth's magnetic field 206–207
ecosystems 50–52, 268
effort 155–157, 268
electric bells 213
electromagnets 208–214, 269
electrons 80, 81, 257, 269
electrostatic attraction 82
endothermic processes 113–118, 269
energy
  in food 28, 31–33, 35
  in food chains 58–60
  resources for 235–238, 240–242
  transfers in endothermic and exothermic processes 113–118
environment 50, 269
estimates 259–260, 269
evaluating an investigation 111, 269
exercise 13, 38, 39
exhaling 12, 269
exoplanets 258, 269
exothermic processes 113–118, 269

F

fats (lipids) 28, 29, 30, 33, 35, 269
feet 164
fibre 28, 29, 269
filters 192–194, 269
fitness 39
floating 152–153
food 28–36
food chains 58–60, 269
food labels 29–30
food webs 65–66, 269
forces 149
  adding 150–152
  balanced and unbalanced 149–154
  effects of 149
  turning forces (moments) 155–160, 271
forensic science 85
fossil fuels 237–238, 269
freezing 113
frequency 53, 269
friction 149–150, 154
fuel cell cars 117
fuels 115, 118, 232–234, 269
  fossil fuels 237–238, 269

G

galaxies 247, 253–260, 269
gas pressure 166–168, 269
gas exchange 9–10, 269
glacial periods 226–227, 231, 269
glaciers 226, 228, 269
Global Positioning System (GPS) 135, 140
global warming 233–234, 269
gradient 141–142, 269
greenhouse gases 230–232, 269
guard cells 9, 269

H

habitats 49–57, 239, 269
haemoglobin 4, 5, 269
hazards 6, 30, 107, 269
health 28, 29, 34, 37–44
heart disease 39, 270
herbivores 58, 270
high blood pressure 39, 270
hinge joints 19, 270
human activities 232–234
hydraulic devices 169
hydrocarbons 233, 270
hydroelectric power 236, 240, 241, 270
hypotheses 41, 154, 228, 270

I

ice age cycle 226, 270
ice ages 226–227, 270
ice cores 227, 228, 270
ice sheets 226, 227, 228, 270
illegal drugs 40, 270
images 179, 191, 213
incident ray 180–182, 270
independent variable 6, 32, 33, 212, 270
industrial lifting magnets 212
inhaling 12, 270
injuries 21
insoluble substances 90, 270
intercostal muscles 11, 12, 270
interglacial periods 226, 227, 270
interstellar dust 256–257, 270
interstellar gas 256–257, 270
invasive species 64–68, 270

J

joints 17, 18–19, 270

K

knives 164

L

law of reflection 181
lenses 188
levers 155–156
lifestyle 37–44
light 178–195
  coloured light 189–195
  reflection 179–183
  refraction 184–188, 272
light gates 138–139, 140
light-years 255–256, 270
line graphs 41–42, 44, 270
line of best fit 41, 180, 181, 270
liquids, pressure in 168–169
load 156, 157, 160, 270

lung volume 15–16
lungs 9–12, 270

M
maglev trains 213
magnetic fields 200, 202–207, 270
magnetic force 200, 201, 271
magnetic materials 202–203
magnetic poles 201, 202, 271
magnetising 202, 203, 271
magnetism 199, 200–202, 271
magnets 199–202
  bar magnets 201, 202, 204–205, 206
  electromagnets 208–214, 269
  field between two magnets 206
  permanent magnets 201, 271
melting 113–114
mercury 61–62, 104
metals 106, 119–125
Milky Way 247, 253–255, 259–260, 271
minerals 28, 29, 271
mirrors 179–183
mixtures 83, 85, 105–108, 271
models 61, 63, 79, 271
  food chains 58–59
  food webs 65–66
  of atoms 79–82
  of diffusion 171
  of dissolving 91–92, 95
  of galaxies 259–260
  of gas pressure 167–168
  respiratory system 14–15
moments (turning forces) 155–160, 271
MRI scanners 199, 213
mucus 11, 271
muscles 12–13, 19–20

N
nails 164
naming compounds 101–102
native species 66, 271
neutralisation 107, 108, 115, 271
neutrons 81, 271
nicotine 41, 271
non-native species 66, 271
non-renewable resources 117, 236–240, 271
normal 180–181, 185–186, 271
normal force 153, 271
nucleus 81–82, 271
nutrients 4, 28–30, 271

O
obesity 38, 271
omnivores 58, 271
organ systems 17, 271
oxidation 101, 271
oxygen 101–102, 122–123

P
paper chromatography 83–86
Periodic Table 266
periscopes 181–182
permanent magnets 201, 271
pharmaceutical drugs 40, 43, 271
photosynthesis 116, 117
physical factors 49–50, 271
pins 164
pivot 155–160, 271
plane mirrors 179–183, 271
planetary systems 249, 256–258, 271
plasma 4, 271
plastic materials 239–240, 242, 272
plum pudding model 80, 272
population 53, 272
precision 13, 111, 133, 272
pressure 272
  air pressure 169
  on an area 161–165
  in gases 166–168
  in liquids 168–169
primary colours 191, 193, 272
primary consumers 58–60, 272
prism 189, 272
producers 58–60, 272
products 100–101, 107, 108, 272
  waste products 108
protection of species 54
proteins 28, 29, 30, 272
protons 81, 272
pulse rate 6
pure substances 105–106, 272
purity 106, 108, 272

R
range 259, 272
reactants 100–101, 272
reactivity 119–125, 272
reactivity series 119, 272
real depth 186–187, 272
red blood cells 4, 5, 272
reflected ray 180–181, 272
reflection 179–183
refraction 184–188, 272
relaxation of muscles 12, 19, 272
reliability 6, 13–14, 133, 154, 272
renewable resources 236, 240–243, 272
repulsion 201, 272
resources
  Earth's 222–243
  organisms' 49, 64–65, 272
respiration 4, 8–9, 272
respiratory system 9–16, 272
resultant force 150, 152, 272
ribs 11, 18, 273
risk 6, 30–31, 107, 273
risk assessment 30, 107, 273

# Index


robots 20–21
Rutherford nuclear model 80–82

S
salts 94, 107, 273
samples 52–57, 273
saturation point 92
scattering 181, 194, 273
secondary colours 191, 273
secondary consumers 58–60, 273
sinking 152–153
skeletal system (skeleton) 17–18, 273
skull 18, 273
smoking 40–44
sodium 103
solar power 182–183, 240, 241–242, 273
solubility 94–96, 105, 273
soluble substances 83–84, 90, 273
solute 90, 91, 95, 105, 273
solutions 89, 90–93, 105, 273
    concentration 90–92
solvents 83–84, 90–91, 273
spectrum 189–190
speed 136–140, 273
    from distance/time graphs 141–143
starch 28, 31
stars 247, 253–260
stomata 9, 273
streamlining cars 152
sugars 28, 31, 35, 273

T
tar 41, 273
temperature
    effect on solubility 94–95
    measuring changes in 110–112
thermal energy 113, 273
thermit reaction 116
time, measuring 132, 134, 136–137
trachea 11, 273
tracking migrating animals 140
trophic levels 58–60, 273
tumours 41, 273
turning forces (moments) 155–160, 271
type 2 diabetes 38, 273
tyres 164

U
unbalanced forces 149–154, 273
Universal indicator 120, 273
upthrust 152–153, 273

V
variables 6, 32, 33, 211, 274
vertebrae 18, 274
vitamins 28, 29, 34, 274

W
waste products 108
water 28, 103, 104, 120–121
weather 223–225, 274
white blood cells 4, 5, 274
woodlice 51–52, 71
word equations 9, 100–104, 274

# Acknowledgements

The publishers wish to thank the following for permission to reproduce photographs. Every effort has been made to trace copyright holders and to obtain their permission for the use of copyright materials. The publishers will gladly receive any information enabling them to rectify any error or omission at the first opportunity.

(t = top, c = centre, b = bottom, r = right, l = left)

p 2 Left Handed Photography/Shutterstock, p 5 Lens Hitam/Shutterstock, p 9 Ianxztan/Shutterstock, p 10 Vladislav Gajic/Shutterstock, p 18tl Alex Mit/Shutterstock, p 18tr koya979/Shutterstock, p 18b Puwadol Jaturawutthichai/Shutterstock, p 20 Nestor Rizhniak/Shutterstock, p 21 Martin Vorel/Shutterstock, p 27 Golden Rice Humanitarian Board www.goldenrice.org, p 29 Celso Pupos/Shutterstock, p 31t Robyn Mackenzie/Shutterstock, p 31b Oleksandra Naumenko/Shutterstock, p 33t Tatjana Baibakova/Shutterstock, p 33b Oleksandra Naumenko/Shutterstock, p 34 margouillat photo/Shutterstock, p 38 sirtravelalot/Shutterstock, p 39 Maniki_rus/Shutterstock, p 40t Prostock-studio/Shutterstock, p 40b Mediscan/Alamy Stock Photo, p 48 reptiles4all/Shutterstock, p 50l Jürgen Feuerer / Alamy Stock Photo, p 50r Lenka Cermak/Shutterstock, p 51 Henrik Larsson/Shutterstock, p 52l Nigel Cattlin/Alamy Stock Photo, p 52r ephotocorp / Alamy Stock Photo, p 53t Hunter Kauffman/Shutterstock, p 53cl David Crausby / Alamy Stock Photo, p 53cr Jack Barr/Alamy Stock Photo, p 53b Polina Tomtosova/Shutterstock, p 54 aleksandr shepitko/Shutterstock, p 56t Mingfei Hou/Shutterstock, p 56tc BLUR LIFE 1975/Shutterstock, p 56bc The Natural History Museum/Alamy Stock Photo, p 56b Cosmin Manci/Shutterstock, p 59l & p60l Aggie 11/Shutterstock, p 59c & p60c Anan Kaewkhammul/Shutterstock, p 59r & p60b Eric Isselee/Shutterstock, p 62t Harry Collins Photography/Shutterstock, p 62b Reproduced with permission from the California Office of Environmental Health Hazard Assessment, p 65t Imran Ashraf/Shutterstock, p 65tl Soru Epotok/Shutterstock, p 65tcl Eric Isselee/Shutterstock, p 65tcr nattanan726/Shutterstock, p 65tr Lucian Coman/Shutterstock, p 65cl volkova natalia/Shutterstock, p 65ccl Isabel2016/Shutterstock, p 65ccr Poppap pongsakorn/Shutterstock, p 65cr Yerbolat Shadrakhov/Shutterstock, p 65bl LizCoughlan/Shutterstock, p 65bc Protasov AN/Shutterstock, p 65br Lee Foster / Alamy Stock Photo, p 67t Emiliano Pane/Shutterstock, p 67bl Giedriius/Shutterstock, p 67br seawhisper/Shutterstock, p 68l Deborah Benbrook/Shutterstock, p 68r Tim Mainiero/Shutterstock, p 71t Henrik Larsson/Shutterstock, p 71c Sarah2/Shutterstock, p 71b Hhelene/Shutterstock, p 76tl Natalia Kuzmina/Shutterstock, p 76tr Adalbert Dragon/Shutterstock, p 76cl Natalia Kuzmina/Shutterstock, p 76cc Pablo Jacinto Yoder/Shutterstock, p 76cr Alejo Miranda/Shutterstock, p 76bl ingehogenbijl/Shutterstock, p 76bc MO_SES Premium/Shutterstock, p 76br robert_s/Shutterstock, p 78 amfroey/Shutterstock, p 80 Andris Torms/Shutterstock, p 84t Nasky/Shutterstock, p 84b Elena Sherengovskaya/Shutterstock, p 85 Pablo Paul / Alamy Stock Photo, p 89 C.W.P.S studio/Shutterstock, p 90 MARTYN F. CHILLMAID / SCIENCE PHOTO LIBRARY, p 94 galichstudio/Shutterstock, p 95 Cagkan Sayin/Shutterstock, p 99 Dana.S/Shutterstock, p 100tr Yakubovich Vadzim/Shutterstock, p 100bl Victoria Lipov/Shutterstock, p 100br Kapuska/Shutterstock, p 103t ANDREW LAMBERT PHOTOGRAPHY/SCIENCE PHOTO LIBRARY, p 103b GIPhotoStock/SCIENCE PHOTO LIBRARY, p 104 Kaentian Street/Shutterstock, p 105 Konstantin Chagin/Shutterstock, p 106 dencg/Shutterstock, p 108t Craig Walton/Shutterstock, p 108b TORWAISTUDIO/Shutterstock, p 109 Nosov Dmitry/Shutterstock, p 110 studiomode/Alamy Stock Photo, p 111 Sheila Fitzgerald/Shutterstock, p 114 O_Lypa/Shutterstock, p 115t Marina Lohrbach/Shutterstock, p 115bl Ramona Edwards/Shutterstock, p 115br Jag_cz/Shutterstock, p 116t Kapuska/Shutterstock, p 116b IgorZh/Shutterstock, p 117 Takashi Images/Shutterstock, p 118 Fouad A. Saad/Shutterstock, p 120t Red Moccasin/Shutterstock, p 120b sciencephotos/Alamy Stock Photo, p 122t TREVOR CLIFFORD

# Acknowledgements

PHOTOGRAPHY / SCIENCE PHOTO LIBRARY, p 122b historiasperiodicas/Shutterstock, p 123t Mimadeo/Shutterstock, p 123b bluehand/Shutterstock, p 132 Adam Vilimek/Shutterstock, p 134l Zack Frank/Shutterstock, p 134cl Garsya/Shutterstock, p 134cr Mircea Maties/ Shutterstock, p 134r pogonici/Shutterstock, p 135 pilotv/Shutterstock, p 138 Bryon Palmer/ Shutterstock, p 140 Reproduced by permission of RSPB, © 2019 All rights reserved; Source: http://www.rspb.org.uk, p 148 Alexander Rochau / Alamy Stock Photo, p 152l Philip Lange/ Shutterstock, p 152r nitinut380/Shutterstock, p 154 Dreamsquare/Shutterstock, p 155 Mark Herreid/Shutterstock, p 156t bluehand/Shutterstock, p 156c NoPainNoGain/Shutterstock, p 156b Andrey_Popov/Shutterstock, p 159 zendograph/Shutterstock, p 163 Independence_ Project/Shutterstock, p 164t eAlisa/Shutterstock, p 164bl Jason O. Watson Photography / Cycling / Alamy Stock Photo, p 164br maxpro/Shutterstock, p 166 Beth Van Trees/ Shutterstock, p 168 GIPHOTOSTOCK / SCIENCE PHOTO LIBRARY, p 172t charistoone-travel / Alamy Stock Photo, p 172b ANDREW LAMBERT PHOTOGRAPHY / SCIENCE PHOTO LIBRARY, p 175 ChiccoDodiFC/Shutterstock, p 177 robuart/Shutterstock, p 178 wacomka/Shutterstock, p 183l Jim West/Alamy Stock Photo, p 183r Kit Leong/Shutterstock, p 184 Jan van der Hoeven/Shutterstock, p 188 Rawpixel.com/Shutterstock, p 190 GIPhotoStock X/Alamy Stock Photo, p 191 & p 198 eightstock/Shutterstock, p 195l LuckyBall/Shutterstock, p 195c Gal Istvan Gal/Shutterstock, p 195r HSU CHIEN PING/Shutterstock, p 199 SpeedKingz/ Shutterstock, p 200 Nata-Lia/Shutterstock, p 201 GraphicsRF/Shutterstock, p 202l sciencephotos / Alamy Stock Photo, p 202r revers/Shutterstock, p 204 mohd kamarul hafiz/Shutterstock, p 205 charistoone-images / Alamy Stock Photo, p 206 Soleil Nordic/ Shutterstock, p 207 ViewStock/Shutterstock, p 209t sciencephotos/Alamy Stock Photo, p 209b Zig Zag Mountain Art/Shutterstock, p 212t Petar An/Shuttterstock, p 212b Anton Kozlovsky/ Shutterstock, p 213 QOcreative/Shutterstock, p 222 AngleStudio/Shutterstock, p 224tr Anneka/Shutterstock, p 224tl bogadeva1983/Shutterstock, p 224b Harvepino/Shutterstock, p 227l SOURADIP HALDER/Shutterstock, p 227c Dino Fracchia / Alamy Stock Photo, p 227r Nick Fox/Shutterstock, p 231 Ziablik/Shutterstock, p 232 Neil Mitchell/Shutterstock, p 233t Asianet-Pakistan/Shutterstock, p 233b Creativa Images/Shutterstock, p 236t Isarapic/ Shutterstock, p 236b Alessandro Colle/Shutterstock, p 237 Anticiclo/Shutterstock, p 239t Johanna Veldstra/Shutterstock, p 239b Science History Images / Alamy Stock Photo, p 240 Joerg Boethling / Alamy Stock Photo, p 242t Sim Creative Art/Shutterstock, p 242b dani3315/ Shutterstock, p 243 Pawarun Chitchirachan/Shutterstock, p 247 Yuri Zvezdny/Shutterstock, p 249tl NASA Image Collection / Alamy Stock Photo, p 249tc NASA / GODDARD / UNIVERSITY OF ARIZONA / SCIENCE PHOTO LIBRARY, p 249tr dpa picture alliance archive / Alamy Stock Photo, p 249br UPI / Alamy Stock Photo, p 251 Zonda/Shutterstock, p 254 Science Photo Library / Alamy Stock Photo, p 257 Pawel Radomski/Shutterstock, p 258 IrinaK/Shutterstock, p 259 Jenn Huls/Shutterstock, p 265 Alex Mit/Shutterstock.